WHO DOWNED THE ACES IN WWI?

NORMAN FRANKS

GRUB STREET · LONDON

Published by
Grub Street
The Basement
10 Chivalry Road
London SW11 1HT

Copyright © 1996 Grub Street, London
Text copyright © Norman Franks

British Library Cataloguing in Publication Data
Franks, Norman L R (Norman Leslie Robert), 1940–
Who downed the aces in WWI?: facts, figures and photos on the fate of over 300 top pilots of the RFC, RNAS, RAF, French and German air services
1. Fighter pilots – Biography   2. World War, 1914–1918 – Aerial operations
I. Title
940.4'4'0922

ISBN 1 898697 51 5

All rights reserved. No part of this publication may be reproduced, stored in a retrieval system, or transmitted in any form or by any means, electronic, mechanical, photocopying, recording, or otherwise, without the prior permission of the copyright owner.

Edited by Daniel Balado-Lopez
Typeset by Pearl Graphics, Hemel Hempstead
Printed and bound in Great Britain by
Biddles Ltd, Guildford and King's Lynn

CONTENTS

Acknowledgements		iv
Introduction		v
Chapter One	1915–1916	1
Chapter Two	January–April 1917	16
Chapter Three	May–June 1917	32
Chapter Four	June–July 1917	46
Chapter Five	July–September 1917	60
Chapter Six	September–November 1917	73
Chapter Seven	November 1917–March 1918	87
Chapter Eight	March–April 1918	106
Chapter Nine	April–May 1918	120
Chapter Ten	May–June 1918	135
Chapter Eleven	July 1918	149
Chapter Twelve	August 1918	163
Chapter Thirteen	September 1918	177
Chapter Fourteen	September–October 1918	190
Chapter Fifteen	October–November 1918	203
Epilogue		211
Index		214

ACKNOWLEDGEMENTS

This book is the culmination of several years of research into the air actions over France and Belgium during World War One, and even of more years of dedicated interest into the lives and careers of the fighter aces of that conflict. Over those years help and advice have come from many sources, some long since ended by 'anno domini'. However, amongst the 'survivors' of my contemporary air historians a few must stand out as having helped or advised me, although I must add that in the final analysis, any conclusions are mine. Some will disagree with them, but that is natural and anything that helps to stimulate further work and research can never be wasted effort.

My thanks go to my friends, Frank 'Bill' Bailey, Russell Guest, Hal Giblin, Stewart Taylor, Stuart Leslie, Keith Rennels, Alex Revell, and Beryl and Derek Winter, for their help, not forgetting the staff of the Public Record Office at Kew.

INTRODUCTION

There is still a vast amount of interest in WWI aviation and particularly in the so-called air 'aces'. Books have been written about any number of them or even by them over the years – von Richthofen, McCudden, Mannock, Guynemer, Udet, Hawker, Fonck, Rickenbacker, Biddle, Boelcke; the list goes on.

The fascination with these first air war fighter pilots seems endless and several who did not survive that far off conflict have their deaths surrounded in mystery or even controversy. Very often while we know when they died, it is not possible to find out how they did so, or who on the opposing side was responsible. Perhaps in the years immediately after WWI this was not particularly desirable, as any number of the victors in these air combats were still alive, but with the passage of time things change. It is now of interest to learn who shot down whom, in these aerial dog fights, in order to help complete our picture of that first air war. There is also the interest in knowing how such and such a 'hero' failed to survive an air battle after being so successful for a period of time. One might believe that it had to be an equally victorious and successful enemy pilot who vanquished someone of equal standing, but, although on occasion this was so, an ace was just as likely to be shot down by a novice pilot as by one of the aces of the other side.

Here for the first time in one volume, are the stories and a record of the aces who fell in combat on the Western Front in WWI, and in the vast majority of cases, who shot them down. Sometimes of course, if it was a new(ish) pilot who went on to achieve great things himself, the reader can accept that his skill was already there and beginning to develop. Others of course, merely ran out of luck at the wrong moment, and in the whirling air battles between the fragile biplanes above the trenches of the Western Front, fell either to a fiery death or had their machines collapse around them.

The aces dealt with in this book are those who fell in action over France, but not those that may have died in accidents or non-air actions. Quite a number have for a long time not been disputed, especially where, for some reason, previous historians or authors have discovered the victorious pilot being associated with a victory over another ace; often through wartime propaganda, for instance, such as Manfred von Richthofen over Lanoe Hawker VC, in November 1916. Many others, however, are not so well known. Often it is a case of studying the loss, by date, time and location, aircraft type and so on, and tying this up with the claims of the opposing side. This is what I have done in a number of cases while pouring over maps of the French countryside, but where I was still uncertain, I have checked my findings and conclusions with other WWI enthusiasts to see if they had thoughts on a certain action. However, the final conclusions are my own. With such a long period of time having elapsed, it may not be possible from this distance to be absolutely sure of some of the conclusions, as occasionally times have just not been noted or records lost, but most have some basis for accuracy. Where there may still be a doubt, I shall say so.

Otherwise, what follows is, for the record, an account of who shot down the aces, while trying to explain a number of the mysteries, or even myths, that have surrounded some of them for around 80 years.

CHAPTER ONE

1915 – 1916

By the very nature of things, there were few aces in the first year of the war. The very term 'ace' came from the French, who like the Germans publicised their more successful pilots. As the war progressed the young airmen of both these sides began to be fêted as super-heroes, recognised in the streets from newspapers, or in the case of the Germans, on postcards. The British officially did not recognise their 'aces', not wanting to acknowledge them as something special so as not to detract from the rank and file airmen, who were completing their daily tasks with just as much vigour, whatever their job, observation, bombing or reconnaissance. This did not stop the public finding out about their airmen, the media of the day ensuring that, but in reality only a few became household names, whereas in France and Germany, most people became aware of such names as Garros, Pègoud, Hawker, Immelmann and Boelcke.

The Germans, of course, chose a higher number of kills for their 'konones' (aces) – ten – but to bring them into line with the Allies and thereby include more, five has gradually become the accepted score for all fighter aces. In due course, it became these fighter pilots – the scouts, or jagdfliegers – that caught the public imagination, duelling in the skies above France, the loser in battle having often to be resigned to a grisly death amidst fire or wreckage. These early air fighters were mostly lone fliers due to the very nature of the early months of the air war. Aircraft in general were few, could often be unreliable, and there was no tactic yet evolved for large formations of aeroplanes. Sometimes two or three might band together, or as time went on perhaps groups of six British or French machines would sally forth to bomb a railway station, or ammunition dump. If they met two or three German aircraft, single-seaters or two-seaters, then the first air battles took place, what legend and history have recorded as 'dog-fights'.

In the beginning there were no real fighter aeroplanes as they became known later. Most machines were two-seaters, whose primary task was one of reconnaissance. Indeed, it was only due to the coming of trench warfare, which ended the cavalry's role of reconnaissance, that the aeroplane became a necessary evil for the generals. The aeroplane could fly over the trenches and barbed wire and observe what the other side was doing. Once this became established, it became necessary for each side to stop these intrusions. Thus fighting in the air began.

The first ace to fall was the Frenchman, Adolphe Celistin Pègoud, then aged 26, and who already had eight years in the military, initially with the 5th Regiment de Chasseurs d'Afrique, and later with the 3rd Regiment d'Artillerie Coloniale. Taken with the new craze of aviation, he learned to fly at his own expense in 1913, even going to the then extreme lengths of jumping from a plane with a parachute in August of that year, followed by the devil-may-care attitude of looping-the-loop the following month. What were things coming to?!

By the time war came in August 1914, Pègoud was already an established and famous pilot, and was immediately assigned to a unit to protect Paris from air assault, but it was in the reconnaissance role that he achieved his first citation, after making a flight over Mauberge on 2 September. From the rank of Brigadier (Corporal) he was promoted to Sergent (Sergeant) and decorated with the Medaille Militaire after tackling his first German aeroplanes on 5 February 1915. In these early fights, Pègoud too was flying a two-seater Morane, although later he progressed to a single-seater. As 1915 advanced he sought further air combat and by mid-July had out fought six German machines and been honoured by being made a Chevalier de la Légion d'Honneur.

Not all were victories in the now established tradition, for in these early days, with air fighting very new, nobody knew how, if, or why to count a score. Later victories, for the French, were enemy aircraft known to have been destroyed, either by crashing, burning, or its pilot falling from the machine. However, in the early days, the French and the British counted what might be said to be a moral victory, i.e. driving off, or driving down, a hostile machine, thereby stopping its crew from doing its work, or forcing a machine to land within its own lines. In the latter case, the reasons for making a landing could be numerous, from being simply out fought and shaken, to being wounded (mortally or otherwise) or just having one's machine damaged. Later these 'forced to land' 'victories' were discounted, even though in some cases, the pilot might be dying and his machine damaged beyond repair.

Finally Pègoud's fighting career was ended on the last day of August 1915. All of his opponents thus far had been two-seater recce machines, four being reported as Aviatiks, another a Taube. Thus it was another two-seater he encountered on this last day, flying a photographic sortie. The pilot of the two-seater was Unteroffizier Kandulski, and by a strange quirk of fate the German had been one of Pègoud's students, instructed in the art of flying when Pègoud had imparted his new found knowledge to others wanting to learn. Kandulski's observer fired on Pègoud's

Morane as it came into the attack, and the Frenchman was hit and fell to his death. He had had a lucky escape from a similar encounter three days earlier, but on this day his luck ran out.

The Fokker period
In late 1915, early 1916, the Fokker Eindekker fighters of the German Air Service appeared. Created by the Dutch aeroplane designer Anthony Fokker, whose talents were recognised by the Germans after he was turned away by both the French and British, his monoplane, fitted with a machine-gun which fired through the blades of the propeller, opened a new chapter in air warfare.

Until now, the various air forces had used several methods of firing guns in the air from their aeroplanes, either over the propeller blades by mounting the gun on the top wing of biplanes, or fixed to fire wide of the blades, the gun being fixed at an angle to the fuselage side by the cockpit, or the ingenious development of the pusher-type aeroplane. With the pushers the engine was mounted behind the cockpit and 'pushed' the aeroplane rather than pulled it. Thus the forward area was totally clear of a propeller, giving the gunner/observer in the front cockpit of the fuselage nacelle a clear field of fire. With the arrival of the Fokker and its forward-firing gun(s), the air war changed.

Like all types, the Fokkers were few in number, and as yet nobody had really come up with the idea of putting the same types of aeroplane in each squadron, escadrille or flieger abteilung. Therefore the new Fokkers were assigned in ones and twos to each flying unit both for protection flights and hunting expeditions. Very soon a handful of German Fokker pilots, the first true fighter pilots, began to put some dominance into the air war over France. Oswald Boelcke, Max Immelmann, Wilhelm Frankl, Kurt Wintgens, Otto Parschau, Gustav Leffers, Rudolf Berthold and Walter Höhndorf were just some of these early victorious pilots.

Immelmann: shot down or gun gear failure?
The first to fall was one of the most famous, whose name is still among the top two or three names known by anyone interested in WWI flying. He and Boelcke had been flying with Flieger Abteilung Nr.62 and assigned Fokkers to fly being deemed to have the required aggressive attitude. They had vied for victories at a time when fighting airmen on both sides of the line considered air fighting a kind of sport; deadly, but a sport nevertheless.

By the end of May 1916, both men had received the coveted Ordre Pour le Mérite, Germany's highest bravery award, and Immelmann had scored 15 victories, Boelcke 18. In the late afternoon of 18 June,

Immelmann had a fight with some FE2b machines – a large pusher type used by the RFC – of No. 25 Squadron RFC, bringing down one at 17.00 hours over Bucquoy. No sooner had he returned from this sortie, his machine refuelled and re-armed, than once more British aircraft were reported over the front. Immelmann, still only 25 years of age, an army cadet since 1905, aged 15, and a flier since late 1914, headed for the enemy along with Unteroffizier Wolfgang Heinemann. Three other pilots, Leutnant Max von Mulzer, Unteroffizier Albert Oesterreicher and Vizefeldwebel Alfred Prehn had taken off shortly beforehand.

Two FEs of 25 Squadron had left their airfield at Lozinghem, west of Bethune, at 19.47 hours, which was 20.47 German time, for a recce sortie to the Lens area. Lieutenant G R McCubbin and his observer, Corporal J H Waller, were in 6346, the other crewed by Lieutenant J R B Savage and 2AM T N O Robinson was 4909. By 21.05 British time, the FEs were in the vicinity of Loos, and the three Fokkers were spotted behind the lines, McCubbin turning towards them. One Fokker dived away while the other two headed towards Lens, they having seen Savage's FE. It seems as if the two German pilots had not yet seen McCubbin, for one was intent on diving on Savage's machine while the other circled slightly above. McCubbin dived after the attacker, Waller opening fire as soon as he was in range. No sooner had he done so than both men saw the Fokker immediately turn to the right and dive perpendicularly towards the ground, seen to crash by the British 22nd AA Battery.

Meantime, John Savage's machine had been hit and the Bradford-born man, not yet 18, had been mortally wounded. The FE continued down and was landed near Lens, where young Savage soon died, the last of a long line of military men. Robinson had also been wounded, but survived as a prisoner of war.

But what of Immelmann? His Fokker had continued down and apparently began to break up, losing its tail and then wings. Ever since that moment, controversy has reigned over how Immelmann met his death. There had been some problems with the interrupter gun gear on some Fokkers, and it was believed that Immelmann had on this evening shot off his own propeller – presumably one blade – which caused enough vibration to tear the machine apart in the air. This was later adopted as the official line by the Germans. However, the British credited the Fokker to McCubbin and Waller, and their combat report seems quite clear on the subject with no suggestion that their fire did not find the target. Was Immelmann himself hit? The Germans make no mention publicly, so one has to wonder, if he had indeed smashed or damaged his propeller, why did he not switch off the engine immediately, which would have stopped the vibration? He could then have glided down to a safe landing.

Some years later, McCubbin, from his home in Germiston, Transvaal, Union of South Africa, wrote to the editor of Popular Flying, an aviation magazine produced in the 1930s and edited by another famous pilot of WWI, W E Johns, the creator of Biggles. McCubbin wrote:

'On the 18th June, Lieutenant Savage with his observer and myself with my observer, were sent up for the last patrol of the day merely to keep an eye on the line between La Bassée and east of Lens. At about nine in the evening, we both saw three Fokkers at the back of Lens.

Savage and I were quite a distance apart but we signalled to each other that we were going to engage these Fokkers. Savage, whilst proceeding towards them, suddenly signalled that he was returning. He was much nearer the Fokkers than I was, and they apparently noticed this as well, and one dived on him immediately. I was flying much higher than they were and immediately dived on the one that by this time was on Savage's tail, but did not open fire. The other two got on my tail, with the result that you had a string of machines all diving down. Savage's machine suddenly got out of control, as the Fokker had been firing at it, and Savage's machine went down. By this time I was very close to the Fokker and he [Immelmann] apparently realised we were on his tail, and he immediately started to do what I expect was the beginning of an "Immelmann turn". As he started the turn we opened fire and the Fokker immediately got out of control and went down to earth.

I then turned to see what the other two machines were doing, who had been firing at me, but found that they had turned and were making back to their own lines, which to my mind rather proved they knew that Immelmann was in the other machine. I went down fairly low to see what had happened to both Savage and the German machine, but as it was getting dark, I could see nothing, and although I flew around for some considerable time I had to give it up and go back to my aerodrome and report the encounter.

With regard to the German statement that Immelmann crashed because he shot his prop off, it is quite on the cards that our bullets not only got him, but his prop as well, and that would be the reason for them trying to make this statement.'

George McCubbin received the DSO for this action, while Corporal Waller was awarded the DCM. Before the month was out, however, McCubbin was badly wounded in the arm, although Waller claimed another Fokker shot down in the fight (07.30, 26 June, still flying 6346).

It appears that von Mulzer took the credit for Savage's FE, the machine having the name 'Baby Mine' on the nacelle. It was to be his fourth victory, and he was photographed next to the disassembled machine the next day. This would have been quite acceptable, for Immelmann was not able to make a claim, and for all he knew, the FE he attacked, which was undoubtedly McCubbin, had been the one later found landed with a wounded crew. As McCubbin noted, by the time the fight was over the other Fokkers were heading east.

Whatever the cause of Immelmann's demise, the first of Germany's great air aces was dead, and the first of many controversies was in place for future generations of air historians to puzzle over.

Less than a month later, the second Fokker ace was to die. East Prussian Leutnant Otto Parschau was 25 years old and was yet another soldier turned airman, learning to fly in 1913 while still with his 151st Infantry Regiment. Within the next year he won many flying awards before war came, whereupon he began flying two-seaters at the front. Because of his aggressive attitude he was soon singled out to fly scouts and was given command of Kampfgeschwader Nr.1, flying Fokkers and later Halberstadt DIs attached to this unit. He gained his first victory over the Argonne on 11 October 1915 and by July 1916, having transferred to command FA32, had brought his score to eight, including two observation balloons. At this time, eight kills was enough to be awarded the Pour le Mérite, Germany's highest gallantry award, which he duly received, but less than two weeks later, on 21 July, he fought his last air battle.

During a combat above Grevillers, just to the west of Bapaume, his fighter was hit and Parschau struck in the head and chest. Mortally wounded he came down but did not survive long. This action, on the British part of the front, was one of many on this day. In fact, the RFC recorded a total of six hostile aircraft brought down with at least three more damaged. Whether Parschau was one of them is uncertain, depending on if any RFC pilot thought any actions by the German pilot were the result of his fire, or merely of a man trying to fly away from trouble.

The time of his last combat is not known, but if the locality is correct then FE2s of 22 Squadron and DH2s of 24 Squadron were in action with enemy aircraft over Bapaume – five LVG two-seaters and two Fokkers. The Germans were scattered in the engagement and a Morane Parasol of 3 Squadron, which had initially been the focal point of the German's attention, and two de Havillands did force one Fokker down from 7,000 feet, seeing it hit the ground near Warlencourt, which is just a couple of kilometres from Grevillers. The two DH pilots were Captain J O Andrews and Sergeant W Piercey, but other fights this day resulted in Fokkers landing in fields near Le Transloy, while Second Lieutenant H C Evans saw a Fokker he engaged go down into the town of Combles. Both these latter places are just south of Grevillers/Bapaume. Evans's victim may have been a two-seater and was in the evening, while Andrews got a Fokker at 08.00 in the morning. Both Andrews and Evans would become aces themselves; 22 Squadron's victories were over Rolands.

Kurt Wintgens
The third noted Fokker ace to fall was Leutnant Kurt Wintgens, aged 22, on 25 September 1916. The son of an army officer, Wintgens became a cadet in 1913 and was attending a military academy when war came, saw some action but then transferred to aviation in late 1914/early 1915, initially as an observer. In March 1915 he trained as a pilot and due to his ability he straight away became a Fokker pilot with FFA67, then FA6b (the 'b' denoting a Bavarian unit, just as 'w' noted a Württemberg unit and an 's' Saxon). He claimed his first victories on 1 and 4 July although they were not confirmed, but his first official kill fell on the 15th after he had moved to FA48. A return to FA6b on the French front saw him raise his score to eight, thereby receiving the Pour le Mérite but then he was assigned to FA23's Kampfeinsitzerkommando (a unit's single-seater fighter unit) Vaux, based on Vaux airfield. With Kek Vaux his score mounted through the summer of 1916 until it stood at 19, with three more unconfirmed. When Kek Vaux became Jasta 4 with the formation of the Jagdstaffeln in August 1916, he was with them but then moved to Jasta 1. By September he was flying on the British front, his final two victories falling on 24 September.

The very next day he was shot down in combat, falling in flames over Villers-Carbonnel. Wintgens had left on patrol at 10.20 (German time – until 25 March 1917 German time was one hour ahead of Allied time), with Leutnant Walter Höhndorf and some other pilots. At 4,000 metres they were attacked by two French aircraft diving with the sun behind them and Wintgen's fuel tank was set on fire. The French ace Alfred Marie Joseph Heurtaux, flying with Spa 3, shot down a Fokker Eindekker at 10.00 at this location, his eighth of an eventual 21 victories. The story is that Wintgens went to the aid of a two-seater, but in defending it he lost his own life. The observer in the two-seater was Josef Veltjens, a future fighter ace and Pour le Mérite winner (35 victories).

Defeat of the Fokker
Mention has already been made of the DH2 fighters of 24 Squadron. It was the DH2, along with the FE2 two-seaters, both pusher types, that

helped defeat the Fokker menace of 1915-16. However, in the period of this bitter struggle for supremacy many pilots were to die, among them some of the early aces. On 3 September 1916, Second Lieutenant H C Evans DSO was killed; with five victories he was among the first British aces.

Henry Evans, despite his low rank, was understood to have been 36 years of age if his 1879 birth date is correct, coming from Surrey, England. He had seen action during the Boer War in South Africa and was living in Canada when the war started. He served in France with the Alberta Dragoons before taking to the air. His rapid achievements with 24 Squadron – five victories in two weeks – had won for him the DSO. However, on 3 September he was shot down and killed flying DH2 No.7887 on a morning Offensive Patrol (OP) over the British 4th Army Front.

A report from the 11th Anti-Aircraft Regiment noted a DH2 being shot down at 1105 while in combat with three enemy aircraft, a shell also seen to explode beneath it. Another member of 24, Lieutenant P A Langan-Byrne also reported seeing a DH2 apparently landing safely north-east of Flers, but Evans had not survived. It is not certain who brought him down – certainly none of the early Jasta pilots made a claim this day so it may well have been German AA fire.

Boelcke's 34th

Patrick Langan-Byrne had also received the DSO for his air fighting prowess with 24 Squadron. (In the early days decorations in this new dimension of the air were slightly higher than later when skills in air fighting became more common, resulting in Military Crosses, rather than Distinguished Service Orders.) An Irishman and former Royal Artillery officer, Langan-Byrne had claimed ten victories between 31 August and 16 October 1916 and had just been given command of 'B' Flight of his unit. However, on the 16th 24 Squadron were in a fight with the new Albatros DI biplanes of Jasta 2, and Langan-Byrne was engaged by the leader of this Jasta, Hauptmann Oswald Boelcke, the former Fokker pilot.

Boelcke was a formidable opponent for any British or French pilot by this time, for the German had by this date amassed a remarkable 33 victories, the 33rd having been claimed earlier on this afternoon. It had been the brain-wave of Boelcke's to group single-seaters into fighting units – Jagdstaffeln – rather than having a few fighters attached to each Flieger Abteilung. They were similar now to the British and French fighter squadrons, but slightly smaller in the overall number of pilots per unit.

The DH2 patrol of 24 Squadron had taken off at 15.15 to again operate over the 4th Army Front by Bapaume. The patrol of four machines was suddenly attacked north of Transloy by what was later described as 12 fast scouts, the Fokkers being rapidly replaced on the front by the new Albatros and Halberstadt biplane fighters. Langan-Byrne was seen to fall out of control, followed by a fighter. He appeared to recover at around 6,000 feet but was last seen disappearing into some ground mist, flying north-east and pursuing the German. Boelcke later wrote:

> '. . . we ran into a squadron of six (sic) Vickers single-seaters south of Bapaume at 5.45 pm. The English leader – with streamers on his machine – came just right for me. I settled him with my first attack; apparently the pilot was killed, for the machine spun down. I watched it down until it crashed about a kilometre east of Beaulencourt and then looked around for a new customer. Meanwhile the others were scrapping with several Englishmen. One of the enemy was actually obliging enough to come quite close to me; I gave him a good shaking up until we got quite low down and then he escaped across the lines by some skilful flying.'

Oswald Boelcke himself did not out live the month of October, colliding with another member of his Jasta on the 28th during a fight with 24 Squadron, and fell to his death. He had achieved a remarkable 40 combat victories, considering it was still only 1916.

Max Lenoir
Among the first French aces was Maxime Albert Lenoir, aged 27, who had learnt to fly pre-war. Flying Caudrons and then Nieuport Scouts with N23, he had scored 11 confirmed victories by September 1916, received the Medaille Militaire, Croix de Guerre with eight Palmes and been made a Chevalier de la Légion d'Honneur. He had been wounded twice in recent weeks, then on 25 October he flew on an artillery support sortie and got into a fight with a German aeroplane. A Vizefeldwebel Schrott claimed a Spad VII on this date, although time, place and the German's unit are unknown.

After Boelcke's death Jasta 2 was commanded by Oberleutnant Stefan Kirmaier, aged 26, another who had transferred to aviation from the infantry. Flying with an artillery observation unit FA(A)203 in 1915-16, he was then attached to Kek Jametz, scoring three kills in July 1916, before taking over Jasta 2, having joined this unit on 5 October. He brought down his 11th victory on 20 November, but two days later the Jasta were again in combat with 24 Squadron's DH2s.

At 1310 over Les Boeufs, south of Bapaume, he was out fought by Captain J O Andrews (5998) and Second Lieutenant Kelvin Crawford (5925) and fell to the ground. It was John Andrews' seventh victory (of an eventual 12, with several more driven down), while for Crawford it was his second of an eventual five victories. As it happens, Kirmaier's Albatros came down on the British side of the front lines, a fairly rare occurrence, and was noted as 'captured', but its pilot was dead, hit in the head by a single bullet.

Richthofen and Hawker
The day after Kirmaier's loss, 24 Squadron lost its leader, Major Lanoe Hawker VC DSO. Hawker had been among the pioneer air fighters of WWI, when fighter aircraft 'per se' were far from a specific type.

Hawker typified the early fliers but was only 26. A former Royal Engineer he was naturally interested in the mechanics of things, so that once having learnt to fly pre-war and then finding himself with No.6 Squadron in France in October 1914, it was natural for him to develop not only his skills but his means of waging war. He had won the DSO for an attack on the Zeppelin airship sheds, and when his squadron received a few Bristol Scout single-seaters, he designed a special mounting for a Lewis machine-gun. Attached to the port side of the fuselage, just by the cockpit, it was fixed to fire at an angle past the propeller blades; therefore, to hit a target, he would have to be flying at an angle away from his line of fire. Nevertheless, he mastered the technique and during the summer of 1915 engaged and accounted for seven enemy machines shot down, either destroyed, or forced to land, one inside British lines. As mentioned earlier, by the nature of all things new, awards to airmen tended to be high in these first months of WWI, which, by the yardstick that airmen were learning each and every day how best to fight in this new dimension, is perhaps as it should be. So Hawker, on 25 July 1915, made three flights, engaged the enemy on each resulting in one falling in flames and one forced down inside British lines, and received the Victoria Cross.

Perhaps here we might mention that the Germans too were going through this period of high awards. For instance their first successful aces, such as Boelcke, Immelmann, Wintgens and so on, all received the coveted Ordre Pour le Mérite after eight victories. Once this figure began to be reached with a little more frequency, the figure was raised to 16 and later still to 20. Later in the war, it even tended in some cases to go well past 20 (even 30) before the 'Blue Max' was recommended, and some didn't receive it even then. Others who might have received it failed to do so due to the abdication of the Kaiser as the war came to an end.

By the end of 1915, Hawker was back in England, Britain's first living ace, if that term had meant anything in these early days. With the increasing air activities over the Western Front, more aircraft and squadrons being needed, and the intensity of air fighting on the increase, Hawker was given the task of forming and leading a new fighting unit – No.24 Squadron, equipped with the new DH2 pusher biplane. He returned to France with his unit, as a Major, in February 1916, and gradually the DH2 began to win the contests with the Fokkers. At this period it was not required for squadron commanders to make war flights. Firstly they had an important job to do in commanding such units, and RFC HQ, unwilling to lose such valuable and usually experienced men, often forbade them crossing the enemy lines. Hawker, however, did fly and often accompanied his squadron patrols, even if he didn't lead them, leaving that task, quite correctly, to his flight commanders.

However, on 23 November 1916, flying with J O Andrews and Lieutenant R H M S Saundby, he flew on a sortie across the lines after lunch. They encountered enemy aircraft north-east of Bapaume and after the initial clash, Andrews saw Hawker diving after one but another was on his tail. Andrews drove this off, but was then attacked himself, had his engine damaged and had to break off and head for the lines. Saundby drove off a German from behind Andrews and then he too headed home. Hawker did not return with them. As we now know, Hawker began a classic duel with one of Germany's up-and-coming young aces who had been guided by the late Oswald Boelcke, a man who would soon become Germany's leading air ace – Manfred von Richthofen. Their battle lasted for some time but Hawker's problem, despite his skill in not letting Richthofen get in any telling shots, was that sooner or later he would have to head for home or risk running out of petrol on the German side of the lines.

A problem the Allied airmen had for most days of the war was that the predominant wind direction was from west to east, thereby helping to push them further and further into German territory, and by the same token making it difficult when the wind was strong, for them to head back to the east. The German pilots were fully aware of their adversaries' problems and took advantage of it. In any prolonged fight the British or French pilot would inevitably have to head away west, giving the German a clear advantage of a tail shot. Or they would often wait to engage the Allied airmen on their way back to the lines, with drying fuel tanks, perhaps short of ammunition, and finding themselves having to fight their way through.

Hawker was faced with this problem now, and finally attempted to break off and scoot for the lines. Richthofen, in his Albatros DII biplane, was on him immediately and although Hawker turned a couple of times to ward off the attacks, he was finally hit in the head as he made another lunge for the lines, and fell dead, crashing at Luisenhof Farm, just south of Bapaume, on the Flers road. It was Richthofen's 11th victory. Soon afterwards the German left Jasta 2 (now called Jasta Boelcke in honour of its dead first leader) to command Jasta 11 – into history.

The two-seaters
It was a policy of the Germans that before being considered a fighter pilot, always supposing one wanted to be one, a flier had to have had a period on two-seaters; in most cases anyway. This, of course, gave the man experience of operational flying as well as air fighting. Many embryo Allied pilots would have benefited from such experience without being thrown in at the deep end after just a comparatively few hours at one or two training schools in England. Many German fighter pilots joined a Jasta following a period at a Jastaschule having achieved one or two victories with some Flieger Abteilung or other, whether achieved while an observer or a pilot. It follows therefore that some two-seater crews would achieve acedom.

One early team was Leutnants Wilhelm 'Willi' Fahlbusch and Hans Rosencrantz, of Kasta 1 (Kampfstaffel or fighting section). By the end of August 1916 they had gained five aerial victories, although only their fifth is recorded, a Martinsyde G100 Elephant of 27 Squadron, which they shot down on the 31st. Less than a week later, on 6 September, they encountered a patrol of 70 Squadron's Sopwith 1½ Strutters out on a long range recce sortie that evening. The three Sopwiths were crewed by Captain W D S Sanday MC/Lieutenant C W Busk (A3431), Lieutenants B P G Beanlands/C A Goode (A1902), and Lieutenants Selby/Thomas (A894).

Having turned west from Busigny, way over to the south-west of Cambrai, a hostile aircraft came into the attack from the west, which was identified as a Roland CII. The sun was in their eyes making it difficult to see the Roland, but Sanday engaged the two-seater and fired about ten rounds, but then his front gun stopped. The Roland came broad-side on and the gunner opened fire at the formation, Busk and Thomas firing across from their rear-gun positions, while Paul Beanlands turned and fired with his front gun too. The Roland pilot then turned to come in astern of the Sopwiths, but Sanday turned too and opened up with his front Vickers gun at close range. He saw the Roland hit, catch fire and go crashing to the ground near Elincourt, to the west of Busigny.

William Sanday, from Liverpool, was 33 years old and would later command 19 Squadron, with whom he would gain his fifth victory flying a Spad having added the DSO to his MC by that time. Paul Beanlands shared in this victory over the Roland. In 1917 he flew with 24 Squadron, brought his score to eight and won the MC, but was to die in a flying accident in May 1918.

The other successful German two-seater observer to fall in 1916 was Oberleutnant Hans Schilling and his pilot Leutnant Rosenbachs of FA22, on 4 December. Schilling was 24 and gained his eight successes with Leutnant Albert Dossenbach between January and November 1916. Schilling and Dossenbach were brought down in a fight on 27 September 1916 with 25 Squadron – Lieutenant V W Harrison and Sergeant L S Court (FE2b 4839) – over Tourmignies and were lucky to survive although both were slightly burnt in the crash-landing. As it happens, the FE crew had also been badly hit, Harrison and Court (the latter becoming an observer ace), being forced to land in British lines. Dossenbach then became a single-seat pilot, bringing his score to 15 by June 1917 but was killed the following month, as we shall read later.

Schilling now teamed up with Rosenbachs, but on 4 December they got into combat with one of France's most aggressive fighter pilots, Charles Nungesser, the 24-year-old Parisian flying with N65. He gained two kills on this day to bring his score to 20. He met Schilling west of Nurlu at midday, shooting the Halberstadt C down after a brief scrap. An hour later he shot down an LVG two-seater east of Lachelle to make it 20, of an eventual 43 victories.

Just over two weeks later, on the 20th, Nungesser shot down the 21-year-old ace Leutnant Kurt Haber over Peronne. Haber, from Upper Silesia, had been a pilot with Bavarian FFA6b in early 1916, scoring two victories in company with his observer, Leutnant Kuhl. He then began to fly Fokker Eindekkers with this unit adding two more kills to his total. When Jasta 15 was formed he joined it and downed his fifth victory on 12 October, a 3 Naval Wing Sopwith 1½ Strutter. He was then transferred to Jasta 3 but became the Frenchman's 21st victory as his machine went down in flames at 15.45 that December afternoon.

Nungesser, despite his high score, was not alone among early French aces. Georges Guynemer achieved 25 victories by the end of 1916, Heurtaux 16 and René Dorme 17.

The fighting Gerry Knight
On the same day as Haber had been brought down, Manfred von Richthofen claimed another British DH2 ace, Captain A G Knight DSO MC, of 29 Squadron. Gerry Knight, although born in England, had lived in Canada for most of his 21 years but returned to join the RFC. After a period on two-seaters with No.4 Squadron he had joined Hawker's 24 Squadron and gained seven victories between June and November 1916,

winning not only the DSO but an MC as well. He had then been made a flight commander with 29 Squadron, added one victory to his tally, but had then fallen in combat with von Richthofen. Again it was a British AA unit that reported seeing a DH2 falling in a spinning nose dive, east of Adinfer Wood while Knight had been leading a five-man OP to Rollencourt and Gommecourt at 09.45 that morning. Adinfer Wood is 12 kilometres south-west of Arras and Knight crashed near Douchy, on the south-east edge.

The last flurry of 1916
No sooner were the Christmas festivities over than more aces were involved in air fighting. On 26 December, Leutnant Hans Karl Müller, flying with Jasta 5, was shot down.

From Saxony, the 24-year-old pilot had already seen much service and action, having operated with FA3, then gone on to single-seaters with Kasta 11 where he gained his first victory, then to Kek Avillers, adding two more, then to Jasta 5 upon its formation. He had just downed his ninth victory but another BE2c gunner hit his machine, wounding Müller in the abdomen. The British crew, of 9 Squadron, were Lieutenant R W P Hall and Second Lieutenant E F W Smith. They were attacked from behind but Smith opened up on the diving Halberstadt and when at about 90 feet range, the German pilot appeared to be hit, dived away and crashed. Müller survived his ordeal but saw no further front-line service, becoming a company test pilot for Siemens-Schuckert following his recovery. Ernest Frederick William Smith, of the 1st Leinster Regiment, attached to the RFC, died of wounds the very next day.

On the 27th, Jasta 1 got into a fight with some FE2b pushers of 11 Squadron. Leutnant Gustav Leffers' Albatros DII was one of four claimed shot down, probably falling to Captain J B Quested and his observer, Lieutenant H J H Dicksee (FE 7666) at 11.20 over Wancourt. The FE crew were then hit themselves, by Vizefeldwebel Wilhelm Cymera, who claimed his victory at 12.20 (i.e. German time one hour ahead). Quested brought his crippled machine down inside British lines, both men surviving. Indeed, Quested, already an ace, counted his victory of the 27th as his seventh, and added one more in early January.

Willy Cymera also became a minor ace but had been lucky to survive this far. Back on 22 August 1916 he had had an encounter with a British-flown Nieuport Scout, piloted by an up-and-coming lad named Albert Ball, flying with 11 Squadron. Cymera was flying a Roland two-seater with Kampfstaffel 1 when Ball attacked. His observer, Leutnant Hans Becker (who had been born in London in February 1894), was mortally wounded, and the Roland fell 6,000 feet, before Cymera gained some measure of control, then crashed into the roof of a house near Vaux. It was Ball's 11th victory. After gaining five victories, Cymera would die in combat on the French front on 9 May 1917.

The 27th also saw the demise of the French ace Adjutant Pierre Augustin François Violet-Marty. A former artillery man he became a pilot, flew Maurice Farman two-seaters, then became a fighter pilot with N57. His five victories included at least two Fokkers and he had won the Medaille Militaire, Croix de Guerre and British Military Medal. He fell in flames over Ornes, circumstances unknown.

CHAPTER TWO

January – April 1917

The impact of the Jastas
The year 1917 saw the establishment of the German fighting Jastas. From the initial dozen formed in the summer of 1916, all had now been equipped with a variety of reasonably good single-seat fighters such as the Albatros DI, DII and the Halberstadt equivalents, together with the LFG Roland DII.

To some extent the winter weather was suppressing all operations in the air on both sides, but the Germans particularly were now eager to test their new units and new aircraft and were just awaiting the arrival of spring. However, there was flying – and losses. Sergent Paul Joannes Sauvage of N38 was the first ace to fall, brought down by an anti-aircraft shell east of La Maisonette at 15.20 on 7 January. He had eight official and at least four unofficial victories flying with N65 and then N38, but at 15.20 his life and career ended. La Maisonette is south-west of Peronne on the German 2 Armee front, and Flakzug 155 claimed a Spad in this location, N38 having a few on strength at this time, despite its designation as being Nieuport-equipped.

Exactly one week later, on the 14th, Sous Lieutenant Andre Jean Delorme of the same escadrille was killed, having made ace status on the fifth. Delorme, 26, came from the Loire region and had been wounded during service in the French infantry shortly after the outbreak of war. Returning to the front he was wounded again but was promoted twice in three days to Adjutant. Transferring to aviation be flew Caudrons with C65 and was decorated with the Medaille Militaire, followed by the Légion d'Honneur, having downed two enemy aircraft, one a Fokker, but was wounded yet again on 31 July. Returning once more he was assigned to N38 to fly fighters bringing his score to five. There were no apparent claims by the Germans on the 14th.

The other Imelmann!
The first German ace to fall in 1917 was Leutnant Hans Imelmann, on 23 January. He was not related to the great Max Immelmann, as the spelling of their names should indicate, but he was one of Boelcke's pilots in Jasta 2. From Hannover, this 19-year-old had gained six victories before

tackling a BE2c of 4 Squadron on this date, but the British observer to Captain J C McMillan, Second Lieutenant Hopkins, got in an accurate burst which hit the youngster's Albatros in the fuel tank. The machine burst into flames and went down to crash near Miraumont at 14.05. John Casely McMillan, formally of the Royal Scots Fusiliers, survived for two more weeks but on 6 February died of wounds received three days earlier from AA fire.*

The 20 Squadron antagonist
Leutnant Walter Göttsch seemed to have a private war with the FE2s of 20 Squadron for a while in early 1917. From Altour, near Hamburg, he had been an NCO pilot with FA33 in 1916 but in late September he became a fighter pilot and was assigned to Jasta 8. He had his first run-in with 20 Squadron on 7 January, downing an FE between Ypres and Kemmel Ridge, then he downed two more on 1 February, one being flown by Flight Sergeant Tom Mottershead DCM, who was to win the Victoria Cross in this action.

Two days later, the FEs got him. At 10.45 that morning, he and six other fighters attacked the FE flown by Second Lieutenant C Gordon-Davis and his observer Captain R M Knowles, over Wervicq. Knowles got in the more accurate burst and watched the Halberstadt Scout go spinning down out of control. They did not see if it crashed, but Göttsch had been wounded and was out of action until April. Gordon-Davis, North Staffs Regt, and Knowles, Norfolk Regt, were both awarded the MC for recent air fights.

Walter Göttsch's first victory after his return was another 20 Squadron FE2 and on 24 April he downed his sixth FE of this unit, and then a seventh on 5 May, and yet another on 31 July. However, 20 Squadron got him again on 29 June; probably the crew of Lieutenants H G E Luchford and W D Kennard (A6516) scored the hits that sent his Albatros spinning earthwards trailing smoke, over Zonnebeke at 13.15 that afternoon. Another German ace went down in this latter action, which will be covered later. The FEs were no real pushovers! Göttsch went on to score 20 victories by April 1918; we shall also read more of him later.

Tit for Tat
Another Jasta 8 pilot who was lucky to escape – at first – was Leutnant Alfred Ulmer, who joined the Jasta as a Vizefeldwebel on its formation. By the latter half of January 1917 he had scored five victories but then met 46 Squadron over Dranoutre on 5 February. Whether or not German

---

*Imelmann had wounded Capt E L Foote of 60 Sqn on 26 October 1916, an ace with five victories.

fighter pilots thought that Allied two-seaters were easy game is uncertain. Even if they understood that any enemy aircraft was dangerous, there may have been a feeling that another victory was coming up. But as so many discovered, even the most obsolete-looking two-seater could bite back.

At this period, 46 Squadron were operating with Nieuport two-seaters and at mid-afternoon on 5 February, Captain G Boumphrey and Captain L Findlay, in A3274, were engaged in artillery observation duty just east of Ypres, at 8,000 feet. Suddenly they were attacked from behind by a Roland Scout, its guns distinctly seen firing through the propeller arc. Findlay was wounded at once but turned to his gun and fired 60 rounds at the aggressor, which scored hits. The Roland went down immediately. However, Boumphrey had also been wounded, twice in the leg and once in the arm. Findlay's wounds were to his back, thought to have been caused by an explosion of six live rounds in the drum of his Lewis gun, which had been hit by what they described as an explosive bullet, in the first attack.

Boumphrey flew home and landed without further problems, and later an AA site reported last seeing the Roland diving steeply east of Zillebeke and was not seen to pull out. Ulmer had been wounded himself and made a forced landing near Dranoutre at 16.45 German time. This was another case of a British unit unknowingly taking revenge on an adversary, for Ulmer had shot down a 46 Squadron Nieuport XII on 26 December 1916. After recovering, he returned to Jasta 8 but was killed in action on 29 June – by 20 Squadron, the same fight in which Göttsch was brought down. Lieutenant H W Joslyn and Private F A Potter (A6415) set his Albatros DV on fire over Hollebeke, Belgium, at 13.15 British time, but on this occasion there was no lucky escape for the 20-year-old. Harry Joslyn, a Canadian from Saskatchewan, and Private F A Potter were both destined to rank among the aces, but Ulmer was their first kill.

The Canadian who went on to command an American pursuit group
Harold Hartney was a Canadian, from Pakenham, Ontario. Born in April 1888 he served in the Saskatoon Fusiliers (Militia) after attending Toronto University, and was a married man when he sailed for England in 1915. Transferring to the RFC he found his way to 20 Squadron after pilot training, aged 28. He operated with this unit until he was wounded on St Valentine's Day 1917, and in that last fight he and his observer claimed two Albatros Scouts over Passchendaele which brought his score to six. Long after the war, Hartney wrote the book *Up and At 'Em* (Cassell & Co.) in which he described the action of 14 February under the chapter title of 'I Meet Mr Richthofen'.

It probably goes without saying that any number of people who were shot down in WWI consider it had to be von Richthofen who got them; even at

JANUARY – APRIL 1917

this stage of the war, the name of von Richthofen was hardly known outside Germany, and probably not at all by a lowly RFC captain. Later in the war, and afterwards, yes! It was similar to German Messerschmitt pilots in 1940 only being shot down by Spitfires, never by Hurricanes!

On this St Valentine's Day, photographs were required of the forest area around Passchendaele. Hartney and his American observer, Second Lieutenant W T Jourdan, with Second Lieutenants F J Taylor and F M Myers MC in a second FE, headed out that afternoon but near Zonnebeke were engaged by enemy fighters and both were brought down. Hartney believed they were shot down by von Richthofen's squadron, Hartney himself by the Baron; or at least that sounded good for his book. However, this was not correct.

Hartney and Taylor were intercepted by Jasta 18, Leutnant Paul Strähle claiming an FE down inside the British lines at Zuidschoote, the German's first of an eventual total of 15 victories. The other FE went to Unteroffizier Flemming near Houthulst Forest, his first of two kills. Von Richthofen did score victories on this date but they were BE2c machines of 2 Squadron, not FEs. Myers was killed but the other three, Hartney, Jourdan and Taylor were all wounded/injured. On recovery, Hartney went back to Canada and somehow managed to transfer to the American US Air Service, eventually commanding the American 27th Pursuit Squadron and, later in 1918, the 1st Pursuit Group.

A Staffelführer falls

Leutnant Hans von Keudell was just weeks away from his 25th birthday on 15 February 1917. An army cadet since 1904 he had served in the 3rd Uhlan Regiment seeing action in France and Poland before transferring to aviation. At first he was a bomber pilot and flew during the Verdun offensive in 1915 and after single-seat training he was with Kek Bertincourt. When the Jastas were formed, he was a founder member of Jasta 1 and between August 1916 and January 1917, he scored 11 victories on the British front. He then became Staffelführer of Jasta 27 on 12 February, claiming his 12th victory on the 15th which may not have been officially confirmed, due to the fact that he fell in this air battle at 16.40 British time.

Operating over the front at this hour were the Nieuport Scouts of No.1 Squadron, flying a patrol between Oosthock and Elverdinghe. Second Lieutenant V H Collins (in A6610) had just been chasing a hostile machine and was returning to the lines at 12,000 feet, but spotted four more enemy aircraft which he attacked from behind. Two of them immediately dived away but he fought one whose fire hit his engine and put it out of action. Collins dived sharply to the left, the German close behind, still firing. Collins desperately pulled up to make a left-hand stalling turn with what speed remained and managed to get half a drum of Lewis into the German machine which staggered away, began to fall out of control and burst into flames.

While this had been going on, a Nieuport two-seater (A229) of No.46 Squadron, crewed by Second Lieutenants S H Pratt and Geoffrey Bryers arrived. They saw the fight in progress and the single-seat Nieuport in trouble, so dived to assist. Both men opened fire when in range, Bryers' Lewis gun then jamming, but they saw the German fighter catch fire and go down, crashing near Vlammertinghe. The burning mass in fact came down inside British lines near Hospital Farm and was given the RFC reference number G.11 for 'captured' aircraft (even if the aircraft was a complete wreck).

A couple of months later, after von Keudell's replacement was wounded, the new leader of Jasta 27 arrived; his name was Oberleutnant Hermann Göring, who had been flying successfully with Jasta 26.

Von Tutschek's first
A number of successful pilots were destined to be brought down as the first victory of an opposing pilot during the war. Sometimes it was the victor's only victory, while for others it was the first of a string of kills. So it proved with Adolf von Tutschek, a 25-year-old Bavarian from Ingolstadt.

Von Tutschek had gone through the not unfamiliar route of army cadet from 1910 with a Bavarian infantry regiment, commissioned, and then when war came he fought in France with another Bavarian infantry unit. Although wounded in battle, his prowess as an infantry officer brought him high rewards, winning the Iron Cross 1st Class, the Bavarian Military Merit Order 4th Class with Swords and then the prized Military Max-Joseph Order which allowed him the use of the title Ritter (Knight).

Recovering, he fought at Verdun but was badly gassed on 26 March 1916 which put him in hospital for several months. Deciding enough was enough he then moved to aviation and, after pilot training, joined a Bavarian two-seater unit, FA6b, at the end of October. Further honours came his way, with the Bavarian Military Merit Order 4th Class with Crown and Swords in January 1917. His aggression had him assigned to a single-seat unit, going to Jasta Boelcke on 25 January. He was to gain three victories with this unit and would later command Jasta 12, then Jagdgeschwader Nr.III, bringing his score to 27 before falling himself as we shall read later. However, on 6 March 1917 he gained his first victory.

There were any number of exotic and colourful names at the turn of the century, and none more so than Maximillian John Jules Gabriel Mare-Montembault, who in 1916 was a scout pilot flying with No.32 Squadron. Abbreviated by his friends to a mere 'Monty', he had gained a modest five victories between September 1916 and this March day. Monty had already survived an encounter with the famous Oswald Boelcke back on 10 October, having been chased along the Bapaume-Albert Road and shot down into British lines, the German ace's 34th victory. On 6 March 1917, he met Jasta 2 again, east of Bapaume in mid-afternoon. Flying DH2 No.7882 providing an escort to a reconnaissance machine, Monty had just shot down a German fighter as bullets ripped through his DH2 and knocked out his engine. Behind him was von Tutschek, the novice pilot but highly determined German soldier.

The time was 16.30 German time, and Monty came down near Beugnatre just north-east of Bapaume. Taken captive, that evening Monty tasted his first meal as a prisoner of war; he would have many more before his release at the Armistice, while his victor had just over a year to live.

Build-up to the Arras Offensive
As the spring weather gradually improved, but only gradually that year, other early aces fell. The Frenchman Maréchal-des-Logis Marcel Hauss of N57, a 25-year-old Parisian, who had five official and one probable victories, was shot down over St Mihiel on 15 February during an attack on a two-seater of FA46 – Unteroffizier Müller and Leutnant Filbig. Vizefeldwebel Friedrich Manschott of Jasta 7, who also operated on the French front, was killed shortly after burning a balloon south of Fort Vaux, Verdun, on 16 March. He had scored 12 victories, which included three observation balloons during the month. No sooner had his third balloon erupted than he engaged four Caudron two-seaters but their gunners got him. The possible action was between the Frenchman and aircraft of Escadrille R210. Two Caudron R4s battled against seven enemy fighters one being thought to be shot down by Caporal Lesée and Lieutenant Debect, although both were wounded in the encounter.

Captain Hubert Wilson Godfrey Jones, aged 25, came from Llandudno. After serving with the Welsh Regiment, winning the MC, he transferred to the RFC and, following his training, joined No.32 Squadron, flying DH2s. From August 1916 to March 1917 he accounted for seven hostile machines and had just taken command of 'C' Flight when he was shot down on 21 March (incorrectly recorded as the 22nd by the Germans). Flying A305, he had led an OP over the 4th Army front, taking off at 15.15. They got into a fight with a two-seater over Roupy, and then

Lieutenant Sidney Cockerall saw his flight commander going down in a spinning nose-dive at 1,000 feet, appearing to be completely out of control. Although wounded and injured in the crash, Jones had come down inside British lines and survived. His victors had been Leutnants W Hilf and W Olsen of FA23. Recovering from his injuries Jones later served at the Central Flying School in England, and then commanded 19 Squadron at the end of the war and into 1919, a post he held again between 1925 and 1928. Walter Olsen, the observer, was killed in action on 3 May 1917, over Guise.

Three days later, on the 24th, Leutnant Renatus Theiller of Jasta 5 was killed in combat with 70 Squadron at 09.00, near Arras. From Mülhausen, the 21-year-old German had been a pre-war pilot, gaining his flying certificate (No.511) in September 1913, the day before his 19th birthday. He flew with FFA44 and FFA25 before becoming a fighter pilot, gaining two victories with these two-seater units. By 22 March 1917 his score had risen to 12, and by strange coincidence he had also scored over Captain H W G Jones of 32 Squadron, whom he had forced to land badly shot up, on 15 February at Miraumont.

Bloody April
Despite late snow and strong winds, the Arras Offensive began on Easter Monday, 9 April 1917, with the Jasta pilots just waiting for the opportunity to have the sky full of Allied aeroplanes.

April had begun badly for the German Air Service, with the loss of no fewer than four aces in the first eight days. Leutnant Alfred Mohr of Jasta 3 had six victories by 19 March but when he tackled the BE2c flown by Lieutenants Douglas Gordon and H E Baker of 12 Squadron on All Fool's Day, his Albatros Scout (2012/16) fell in flames near Arras at 11.50. Coming down inside the British lines it became G.18.

The following day Leutnant Erich König was celebrating his 27th birthday and perhaps hoping for his seventh victory. However, Jasta 2 were in a fight with FE2s of 57 Squadron, losing two of their pilots, one of whom was the birthday boy. His Albatros fell in flames at 09.45 over Wancourt. Two of 57's FEs were downed in this fight, one of the crews possibly getting König before they fell.

The third ace was a Pour le Mérite winner, Leutnant Wilhelm Frankl who was killed on the 8th. From Hamburg, he was 23 and another pre-war pilot having gained his licence in July 1913. At the beginning of the war he was with FA40 in Flanders, then as a Fokker pilot with Kek Vaux. With the Pour le Mérite and nine victories, he went to Jasta 4 in September 1916, rising to command the unit on 1 January 1917. On 6 April he scored four victories, at that time a unique achievement, especially as the first was downed in the early hours of the morning – i.e. a night kill – then three in one hour between 08.50 and 09.55. His 20th and last victory was scored the next evening.

Then, on the 8th, which was Easter Sunday, he was in combat with Bristol Fighters of 48 Squadron, a type new to the Western Front. During this action his Albatros DIII, No.2158/16, was seen to break up, his body falling near Vitry-Sailly, at 14.15. No. 48 Squadron made a claim: Captain D M Tidmarsh/Second Lieutenant C B Holland, and Second Lieutenants O W Berry/F B Goodison, an Albatros out of control over Remy, which is just to the south-west of the area recorded by the Germans. Quite often Frankl's score is noted as only 19, but the one usually missed is his first while with FA40. On 10 May 1915, acting as an observer, he shot down a Voisin with rifle fire!

Victories

Before moving on perhaps it is time to discuss victories in WWI. No.48 Squadron had only been able to claim an 'out of control' victory on 8 April. This is to say, the Albatros was seen to go down in such a way as to suggest it would crash, but the actual crash had not been witnessed. This could be due to low cloud or mist, or the pilots being too busy to watch the falling aircraft all the way down. One sure way of being shot down was to take one's eyes off the battle to watch a falling foe. And obviously in this case, nobody had witnessed the Albatros begin to break up at a lower level.

It was really only in the latter half of 1916 that victory scores became of real importance. Everything about the air war in WWI was new – there had never been one before. Thus there developed some strange interest and obsession in the knowledge that some pilot or other had succeeded in shooting down an opposing aircraft. As time progressed, and a handful of pilots seemed to be able to achieve this remarkable feat on a number of occasions, these pilots drew the attention of the popular press in France and Germany. Not to be out done, the British press too began to try and report successful pilots, although as mentioned earlier, the British policy was not to select any particular pilot out for individual glory at the expense of two-seater crews flying dangerous recce sorties, contact patrols, photography missions or artillery observation tasks.

However, once the interest became established, a method of scoring victories followed. As the British, and to some extent the French, maintained a wholly offensive air war, their aircraft were generally fighting on the German sides of the lines. The Germans, due mainly to their smaller fighting numbers, and then because they preferred to take advantage of the enemy coming to them, were mostly able to confirm any aircraft brought down simply because the wreckage, bodies, captured airmen etc., were in their territory. The British, on the other hand, were only rarely able to bring down an opponent on their side of the lines, thus rules for confirming victories – aircraft shot down – were different.

Obviously an aeroplane seen burning in the air or on the ground, or crashing into the ground, breaking up in the air, and so on, were certain kills, but they still had to be witnessed by a third party, either another airman or someone on the ground, near the front, who could confirm the aircraft's destruction. Because most fights were high up, and it was not wise or even not even possible to follow a spinning opponent down, the British recognised an 'out of control' 'victory', provided someone still saw it fall, and fall in such a way as to say it must have crashed.

This sort of claim must, of course, be very difficult to verify, for even the most ardent air fighter would firmly believe his spinning victim must crash – that is human nature. There was no protective armour in these WWI aeroplanes, and quite often an attacking pilot would see the enemy pilot jerk as if hit by gunfire, and slump forward just as his machine whipped into a spinning nose-dive. Another kill; but the man might only be wounded, even only slightly wounded, or perhaps his 'jerk' had been one of surprise. Having spun several hundred feet down and out of the air battle, the pilot, wounded or not, would often pull out of the dive and head for home, a wiser pilot but not a dead one.

When pilots on both sides realised that in most cases they would not be followed down, once they found themselves in trouble it would be easy to flick over and head down, trying to look as 'dead' as possible, in order to escape a situation. That the British pilots did it, but seemed not to care too much that the Germans might be doing it too, seems odd considering the number of 'out of control' victories they claimed during the war, when most of them must have realised that their victim would be away safely once down to a lower level and out of reach. Nevertheless, everyone claimed 'out of control' victories, and they were taken as part of a fighter pilot's 'score', as were observation balloons destroyed. Thus a pilot might have eight victories, made up of three destroyed, one balloon flamed, and four out of control. That one of his destroyed claims might have also been credited to another pilot who equally believed that 'his' target had crashed, and been witnessed by someone else, didn't get evaluated to the extent that only one pilot would be credited, so our eight-victory man may have only achieved two certain kills and one balloon – total three – as all four of his 'ooc' claims had not crashed, but survived to fight again.

In the early years of the war, a 'driven down and forced to land' victory was also credited. Again one must remember the time and period of the events (the learning curve was still curving!). For the British, to force an opponent down to land was a 'victory' of sorts, for the enemy machine had been stopped from doing its job, whether flying a recce sortie, artillery ranging or even fighting the RFC's two-seaters. And in some cases the pilot or gunner or both might have been wounded, perhaps mortally wounded or killed; the RFC pilot had no way of knowing. Just as he did not know that his recent opponent had not been badly hit, but had perhaps decided that prudence was in this case the better part of valour. The pilot had saved himself, his observer and his aircraft. He could always try again later to carry out his assigned task, once the aggressive Englishman had flown off. A number of the early British aces had a few 'forced to land' (FTL) victories in their score – Albert Ball for instance had nine of his 44 victories classed as FTL. Only by early 1917 did this category not count in a pilot's score, unless perhaps the eager British pilot went down, strafed the downed machine and set it on fire. Yes this happened – the Biggles image was a post-war one!*

The French, almost from the start, decided to credit their fighting pilots only with those enemy aircraft or balloons seen to be definitely destroyed and witnessed. Anything similar to an out of control claim was noted as a 'probable' but was not collected in an overall score. In 1918 when the Americans came to the front, they generally adopted the French system, mainly because they came under French sectors and administration in the beginning.

The Germans, as already mentioned, were far better able to confirm kills due to them falling inside their own territory. If a wreck or captured or dead airmen could not be found, then the victory was just not confirmed. If the British or French machine was seen to come down or land inside Allied lines, then sometimes these would not be confirmed, unless someone else saw them burning on the ground, completely smashed, or they had been seen to be shelled by German artillery and destroyed before they could be spirited away or hidden out of sight. Because the German pilots, more than the British or French, could almost definitely achieve degrees of decorations for a certain number of victories, it was very much in the individual's interest to have his kills confirmed, whereas, a British squadron adjutant might telephone a front-line artillery or anti-aircraft gun unit near the front to ask if they had seen and would confirm one of their pilot's reports of downing a machine at such-and-such a time, over such-and-such a locality; they would not generally pursue it much further. On the other hand, the Germans would religiously pursue any chance to have a kill confirmed.

---

*Biggles (Captain James Bigglesworth DSO MC), the fictitious hero created by W E Johns after the war, depicted as the archetypal air fighter of WWI.

A pilot would travel to the area he thought his opponent had come down, check with local troops and so on. A Jasta adjutant would go too sometimes, or send men to find out all about the claimed kill. Later in the war, with some of the big aces, such as von Richthofen, squadron personnel would be despatched to interview survivors to ensure they could verify they had been brought down in the same way as described by the pilot.

With Wilhelm Frankl's loss, neither Bristol Fighter crew had any way of knowing that their fire, or combined fire, had definitely brought down anyone, let alone one of Germany's leading aces of the day. All they could report was that an Albatros had gone down and been lost to view, either by cloud or the heat of battle. They obviously believed – to what extent we will never know – that it had crashed, and in this instance they were right. However, in the vast majority of cases, the 'out of control' victories or victims were simply getting themselves out of trouble, and mightily glad of the chance to do so!

The Battle of Arras
As fortune would have it, the first ace victim of the April Battle of Arras was the same Captain Tidmarsh who had helped down Frankl. David Mary Tidmarsh was an Irishman, born in Limerick in 1892 and was thus 25 years old on 11 April. After serving with the Irish Rifles he had become a pilot and flown DH2s with Hawker's 24 Squadron in 1916 and won the MC while achieving three victories before being rested. He went out to France again in April 1917 with the newly formed 48 Squadron, flying the new Bristol F2a two-seat fighters, which was quite a change from single-seaters.

The new Squadron did not get off to a good start, and the famous battle with von Richthofen's Jasta 11, which he had commanded since leaving Jasta 2 in January 1917, cost them four aircraft and crews, including the famous Captain William Leefe Robinson VC, who had downed the German airship LZ97 in April 1916. As flight commander with 48, Leefe Robinson led a patrol on 5 April which was shot to pieces. Unfortunately they bunched together for mutual protection whereas later the Bristol Fighter became an extremely efficient two-seat fighter machine in its own right, with the pilot and observer able to see off any number of hostile aircraft when used correctly.

As one of the other flight commanders, Tidmarsh and his observer, Holland, had scored two more victories in early April and in their final fight claimed two more, although Jasta 11 did not lose any pilots but were then knocked out of the air and the war by Leutnant Kurt Wolff, his ninth of an eventual 33 victories. Tidmarsh and Holland were both taken prisoner (A3338), later meeting up with Leefe Robinson who was also a guest of the Germans. Two of the other Bristols were also brought down, one by Karl Schäfer (his 15th), the other by Manfred von Richthofen's younger brother Lothar, his second victory. Schäfer's victim came down south-west of Fresnes, the other two at Manville Farm, to the north of Fresnes, all at around 09.10 to 09.15.

Death by collision
The next day saw Leutnant Adolf Schulte of Jasta 12 die. He had scored eight victories by this date and was part of a force which engaged FE2b machines of 18 Squadron north of Barelle. In the whirling mass of twisting, turning, diving aeroplanes, it was not uncommon for collisions to occur, and one happened on this occasion. Two of the FEs failed to get home, but it was the one crewed by Lieutenant Oswald Thomas Walton, 3rd South Lancs/RFC and 2AM J C Walker (4995) into which Schulte flew, all three men falling to their deaths at Rumaucourt at 10.40.

On the following morning (13th) FE2s of 25 Squadron were active, sending aircraft on a bomb raid to Henin-Liétard. Three took off between 18.05 and 18.20, led by Captain L L Richardson/Second Lieutenant D C Wollen (A6372), with an escort of Nieuport Scouts. They had just dropped their bombs and were turning for the lines when the Nieuports left them. Almost immediately they were attacked from behind by eight Halberstadt and Albatros Scouts. Lieutenant D S Holmes, observer to Lieutenant H E Davies (A6373), was hit in the elbow by a shell splinter but they got back. Corporal L Emsden, observer to Lieutenant R G Malcolm (A6383), fought off three Albatros Scouts one of which he shot down, seen to crash by 'E' (AA) Battery. Richardson was engaged by two Scouts but apparently his lateral controls were shot away as he was last seen drifting south-east towards Fampoux. Later, 'C' Battery observed an FE landing under control east of Fampoux which was later seen on the ground by a recce aircraft the next morning.

Meanwhile, six more FEs of the Squadron had carried out a similar bomb raid but without escort. They had expected to see Nieuports but saw five German fighters instead, which attacked while they were further to the north than the earlier formation. Two FE crews had their engines shot out and failed to recross the lines, although one Albatros was seen to crash near Sallaumines, confirmed by British AA.

Lancelot Lytton Richardson was an Australian from Barraba, New South Wales (not American as sometimes noted) and was either 20 or 21 depending on which birth date is correct, the years 1895 and 1896 being quoted. After pilot training he had joined 20 Squadron on 3 June 1916, so he was an experienced operator by this time despite an early wound, having won the MC and with his observers had accounted for seven enemy aircraft. Despite the report that he seemed to make a good landing, both he and Douglas Charles Wollen were reported killed in action. They had been brought down by Leutnant Hans Klein of Jasta 4 – his seventh of an eventual 22 victories – south of Vimy, his claim timed at 19.10. Jasta 11 had downed the other two FEs, at 19.15 and 19.25, one by Manfred von Richthofen the other by Vizefeldwebel Sebastian Festner.

There was another mid-air collision on the 14th, this time down on the French front. Leutnant Hartmut Baldamus, a 25-year-old from Dresden, had joined the Air Service when war came, serving with FFA20 in March 1915 as a lowly Gefreiter, but nevertheless he gained five victories before being commissioned in September.

Transferring to fighters he flew with Jasta 5 but in November 1916 moved to Jasta 9. He scored steadily and by 14 April his score stood at 17. At around 11.30 he was flying with the Jasta over the front and got into a dog-fight with French Nieuport XVIIs of N37. In the battle he collided with one flown by Caporal Simon, over St Marie-à-Py at 11.40, both men crashing to their deaths. Baldamus was given his 18th victory but he knew nothing of it.

The Navy boys get in on the act
There were two British air arms in WWI, the Royal Flying Corps and the Royal Naval Air Service. As one would imagine, the RNAS was mainly concerned with maritime flying, coastal patrols, coastal reconnaissance, anti-submarine operations, and so on. They also did sterling work in bombing, hunting airships and defending French and Belgian coastal ports.

With the RFC being so hard-pressed on the Army fronts, the Admiralty gave permission for certain naval squadrons to support the RFC in the field, and a number of fighter and bomber units did so, being equipped with the same types as their RFC counterparts. Their support was gladly accepted during the Arras battles and on 23 April 3 Naval Squadron, flying Sopwith Pups, shot down the leader of Jasta 12, Hauptmann Paul Henning Aldabert Theodor von Osterroht. Von Osterroht came from Luneberg, the son of a Captain of Dragoons. He was older than most, being 29 at the date of his death, but he had joined the army in 1906. He became a pre-war aviator, gaining his licence in 1912 and in 1914 was with BAO (Brieftauben Abteilung Ostende – a bombing unit hiding under this innocent title of Carrier Pigeon unit, Ostend). One of his observers with this unit had been none other than Manfred von Richthofen.

Von Osterroht then served with Kampfstaffel Nr.1 in 1916 before being appointed Staffelführer of Jasta 12 on 12 October. In this capacity he gained seven victories in early 1917, two being over Naval fighters, a Sopwith Triplane of 1 Naval, and a 3 Naval Pup.

On 23 April, the day he gained his seventh and final victory, he was leading his Jasta again that evening in the vicinity of Cambrai. The Pups of 3 Naval were out too, in force, and among the patrol were some experienced as well as some up-and-coming pilots. In fact most of them would be aces before too long. They were Lloyd Breadner, H G 'Tiny' Travers, G B Anderson, Harold S Kerby, W E Orchard, Francis D Casey, Harold F Beamish, Hubert S Broad, Jack J Malone, Fred C Armstrong, Art T Whealy, Alfred W Carter, S L Bennett and Joe S T Fall.

Near Ecoust St Mein, Jasta 12 ran into FE2s of 18 Squadron and the escorting Naval Pups at just on 18.00. There were several Halberstadt and Albatros Scouts involved, and the problems associated with these whirling WWI dog fights occurred here. As far as can be ascertained, Jasta 33 were also involved, but only von Osterroht was killed, while one Jasta 33 pilot was wounded. Unfortunately, pure losses of aeroplanes aren't always known, so there may have been times when German pilots crashed without serious injury to themselves.

In this air fight, Lloyd Breadner (N6181) claimed two enemy fighters, one falling while leaving a smoke trail and another out of control, into which several other pilots fired. Casey (N6182) attacked an Albatros being engaged by Art Whealy (N6194), firing 40 rounds at point-blank range and it went into a spin; Whealy thought Casey had hit the pilot. Casey also claimed a second Scout out of control. Carter (N6179) fired twice into another Albatros, seeing a tracer bullet of his second burst hit the pilot. The Scout spun then nose-dived, levelled out above the ground and crashed trying to land. Joe Fall (N6203) witnessed this and also claimed an Albatros out of control himself. Kerby (N6160) climbed after two Halberstadts but was seen by one enemy pilot who tried to warn his companion by crossing over him. While doing so the two Scouts collided, the wheels of the one smashing into the top wing of the other; both machines fell away, recorded as '. . . went down crashing through the sky'. Armstrong (N6178) fired 125 rounds into another Albatros and claimed an out of control victory.

From this confused action it is almost impossible to say who might have actually shot down von Osterroht or the Jasta 33 pilot, and it is not too clear if the two machines that collided actually crashed. The Squadron records seem to indicate one destroyed by Carter, two destroyed(?) by collision, credited to Kerby, with another seven out of control. Carter's was the only one seen to crash, and with the pilot appearing to be hit by a tracer bullet, that might well have been von Osterroht. Breadner might have wounded the Jasta 33 pilot, or so could Casey – or any of them!

A Richthofen man falls

On 25 April, Jasta 11 lost one of its up-and-coming ace NCO pilots, Vizefeldwebel Sebastian Festner. A 22-year-old Bavarian, Festner had started the war as a mechanic with FA1 and then served with FA7b. He learned to fly unofficially, then took formal training, following which he flew with FA18 and FA5b. Assigned to Jasta 11 he scored 11 victories in the early months of 1917.

Once again there is some controversy over his final moments, as some reports say his propeller blade broke (but like Immelmann, surely he would have switched off and glided down?) or he was hit by AA fire. The Germans record he was in combat with four BE2s, which appear to be the flight of Sopwith 1½ Strutters he was indeed engaged with.

Six aircraft of No.43 Squadron were on a Line Patrol at 6,000 feet in the vicinity of Oppy at 08.15, having taken off at 07.00. They were attacked by four hostile machines which were described as 'German Nieuports, painted red'. Because the newest Albatros DIII and DVs had a smaller lower wing, similar to the Nieuport Scouts' unusual sesquiplane feature, RFC crews often referred to these machines as German Nieuports (later they sometimes referred to them as V-Strutters, due to the unusual V-strut connecting the wings). By this time, of course, von Richthofen's Jasta 11 machines were virtually all marked with red in some form or another, each individual pilot having another additional colouring as a personal identification, but to the RFC, the aircraft were simply red.

The Germans timed the fight at 09.10, and still being an hour later than British time, this coincides pretty well with 43's time, and their quoted area – north-west of Gavrelle – ties up with Oppy, both locations being just to the north-east of Arras. One of the Strutters was crewed by Lieutenants C R O'Brien, King's Own Royal Lancs Regt and RFC, and J L Dickson, 8th Reserve Canadian Battn/RFC (A8232). East of Oppy they were attacked, one of the four Albatros Scouts coming in on their tail. O'Brien stalled his machine and spun away out of trouble. As he levelled out he saw another Sopwith ahead of him with a German on its tail (Second Lieutenants C L Veitch/E S W Longton A7799).

O'Brien attacked the German with his single front Vickers gun, seeing his tracer hit the Albatros which immediately went into a spinning nose-dive. With the other three Germans coming down at them, O'Brien had no opportunity to see what happened to his Albatros, but later an AA battery confirmed it had crashed. O'Brien had been with 43 Squadron for six months, so was fairly experienced. Festner so nearly got his 12th victory, and had in fact wounded Longton the observer, but had failed to clear his tail, giving O'Brien the opportunity to fire at him. In the deadly world of air fighting, it only took a moment's carelessness to cost a pilot his life.

The next day saw the Frenchman Capitaine René Doumer killed in combat. Doumer was 29 and came from Laon, and had entered military service in October 1908, serving with the 2nd Bataillon de Chasseurs à Pied. By the time war came he was an infantry lieutenant and in the second month of the war he was made a Chevalier de la Légion d'Honneur, although wounded too. Entering aviation he became a pilot at the beginning of 1916, flew Caudron two-seaters with C64 and in March he and his observer had shot down two enemy planes. In September he was given command of N76 (which then became Spa67 when fully equipped with Spad VIIs) and brought his score to seven by March 1917.

In combat on 26 April he met up with the leader of Jasta 19, Oberleutnant Erich Hahn, from Leipzig. Hahn had scored his first victory with Jasta 1 the previous autumn, and coincidentally had shot down two Caudrons in early 1917. Over Brimont he sent Doumer down after a brisk air fight thereby gaining his fourth kill; he would raise this to six by the end of April with two balloons on the 30th.

Doumer's father Paul, a senator and later President of the République, was assassinated in Paris in 1932. The bridge that crossed the Red River on the edge of Hanoi, Vietnam, was named the Paul Doumer Bridge in his honour. In the Vietnam war, this bridge became a major target for American airmen, flying aircraft that WWI airmen could hardly even have imagined as they battled in the skies over the Western Front.

Jack Malone's last fight
On the final day of April 1917, 3 Naval Squadron lost one of its most aggressive pilots, Flight Sub Lieutenant John Joseph 'Jack' Malone. This 22-year-old Canadian from Saskatchewan joined 3 Naval Wing in 1916 operating with 1½ Strutters. Then he moved on to Pups with 3 Naval Squadron as it was forming. During March and April 1917 he achieved a remarkable ten victories and was awarded the DSO. He was brought down flying Pup N6175 on the evening of the 30th at Rumaucourt, northwest of Cambrai, by Leutnant Paul Billik of Jasta 12. Oddly enough it was Billik's first success since joining the Jasta on 26 March. Billik went on to score 31 victories, including a number of aces as we shall read later.

CHAPTER THREE

May – June 1917

Loss of the boy ace
Early May 1917 saw the loss of one of the early heroes of the RFC both at home and in the Service itself. Captain Albert Ball DSO and two Bars, MC was only 20 years old but in the final year of his life had achieved a phenomenal record in air combat with courage and determination, as well as capturing the imagination of the general public in Britain and in other parts of the Empire.

Born in Nottingham in August 1896, he had joined the Sherwood Foresters in 1914, then transferred to the RFC. By February 1916 he was in France, flying with 13 Squadron on BEs. On 7 May he joined 11 Squadron, exactly one year before he was to die. His fighting spirit in the air soon singled him out to fly one of the new Nieuport single-seat fighting aeroplanes, although his first victory was achieved in a Bristol Scout. From then on he was a devotée of the nimble Nieuport with its wing-mounted Lewis gun. Often he would hunt his opponents alone, get himself into a position below his victim, pull down the Lewis and fire up into the generally unsuspecting adversary.

By the end of the summer he had achieved 31 victories of the period. Back home he was an instructor, but then became a flight commander with the newly formed 56 Squadron, the unit destined to take the new SE5 scout to France in April 1917. Flying the new fighter, and occasionally a personal Nieuport, he brought his score to at least 44 by 6 May.

The following evening a patrol of Squadron strength (11 machines from all three flights) flew out at 17.30 to the Cambrai-Bapaume Road area, led by Ball (A4850). One C Flight machine turned back with engine trouble. B and C Flights climbed to 9,000 feet, leaving Ball and another pilot at 7,000 feet (his third man had become separated) the whole force still edging towards Cambrai. Over Bourlon Wood, the evening sky darkening, B Flight briefly entered some cloud, but another pilot became separated. He may have spotted a German and dived away, but in any event he was never seen again, picked off by a Jasta 3 pilot. C Flight then spotted enemy aircraft, Captain H 'Duke' Meintjes leading his pilots down. One of the enemy was hit and fell away, the SEs regaining height, only to be attacked themselves by four red Albatros Scouts – Jasta 11.

MAY – JUNE 1917

Manfred von Richthofen, his score at 52 by the end of Bloody April, was in Germany on leave, Jasta 11 being in the temporary command of his brother Lothar. The SEs engaged the red machines and in a whirling dog fight one Albatros was sent down to a crash-landing, its pilot seriously injured. Meanwhile, Ball and his lone companion had climbed up into the action, Ball attacking one Albatros but then breaking away with assumed gun trouble. His companion's Vickers gun also jammed; then seeing a C Flight pilot being hard pressed, he dived to help. There appeared to be some more Albatros Scouts about, but then some Bristols and Nieuports arrived and the fight broke up.

In the evening sky, the SEs paired up, Ball and his earlier companion, Cyril Crowe, climbing for height. The pilots had several scraps, Meintjes being hit and wounded in the wrist, although he managed to land on the right side of the lines to have his injury attended to. By this time it was 19.00 and the weather was deteriorating, with storm clouds mounting, making the early evening seem darker than usual.

Captain Cyril Crowe, B Flight, met Ball above Fresnoy, north-east of Arras, the fighting having drifted the aircraft north to a position where Ball had said the pilots should rendezvous if separated. Crowe saw Ball signal with two red Very lights, the warning of enemy aircraft, but could not see any, so merely followed Ball; then near Loos, Ball went down on a lone red Albatros. Crowe also attacked, and then Ball again, while Crowe noticed nearby a Sopwith Triplane and a Spad (machines of 8 Naval and No.19 Squadron). Crowe watched below as Ball and the Albatros dog fought each other down into some cloud and lost sight of them. Soon he turned for home with petrol tanks almost empty. Ball continued to pursue the Albatros over Seclin, westwards towards Annoeullin village, and in a desperate attempt to deter his attacker, the German pilot turned to engage, but his machine was hit again, and with a ruptured fuel tank, the German pilot made a crash-landing. The pilot was Lothar von Richthofen.

In those last few moments Ball chased the Albatros into a huge storm cloud. The next people to see Ball were four German officers on the ground some short distance away. Through binoculars, on hearing a stuttering aero engine, one of the men then saw an SE5 come out of the cloud – its engine now stopped, inverted, trailing a thin plume of smoke – go behind some trees still upside down, and crash. It had hit the ground near an abandoned farmhouse named Fashoda, a little over a mile from Annoeullin. Hurrying to the scene, the four men found Ball had been taken from the wreckage by a local French girl. Ball was barely alive, opened his eyes once and, cradled in the girl's arms, passed away. He had suffered a broken back and leg.

The controversy that now followed centred around the fact that the Germans credited Lothar von Richthofen with the victory over the great British ace. However, everything seemed to point to the fact that Ball had died of injuries sustained in the crash. Lothar reported having been in combat with Triplanes, and we know 8 Naval's Triplanes had a fight in the area – Crowe had seen one too – but although the SE5 was still new to the front, the difference between a biplane SE5 and a three-winged Triplane is basic and obvious. It seems fairly certain now that Ball had simply come out of the cloud at low level, inverted, and crashed before being able to correct his machine. It has been suggested he may have become disorientated in the cloud and because of this, his engine carburetor had flooded causing the motor to splutter to a stop. He had been airborne for nearly three hours and been in several fights, so he was obviously more than a little tired, physically and mentally.

In the event, Richthofen was credited with downing Ball, but whether this was little more than a ploy of the German High Command to credit Lothar rather than credit nobody, will never be satisfactorily known. He had been fighting Ball, or at least had been chased and attacked by him, and brought down. Quite how the word Triplane got mixed up in the proceedings is equally obscure. Why didn't the Germans merely report a single-seater (an einsitzer), or just an enemy aircraft? Even calling it a Sopwith would have been acceptable; Germans often mistook aircraft types, and with the SE5 being new(ish), mixing it with a Pup would have been more acceptable than with a Triplane.

Ball was buried at Annoeullin on 9 May, and in June came the announcement that Britain's greatest fighting pilot had been awarded a posthumous Victoria Cross.

Lothar again!
Two days after being brought down by Ball (the 9th), Lothar von Richthofen was again leading aircraft of Jasta 11, this time against the Bristol Fighters of 48 Squadron.

The British machines took off at 16.25 to fly an OP to Arras and in an area east of Vitry and Fampoux, roughly five to six miles down the Scarpe River, Jasta 11 engaged them – possibly Jasta 4 too. One Bristol crew claimed three Albatros Scouts out of control east of Vitry, while Second Lieutenant William Thomas Price and his observer, Lieutenant C G Claye 5th Sherwood Foresters/RFC, destroyed one between Vitry and Fampoux.

Price had transferred to the RFC from the 13th Royal Warwickshire Regiment and had been one of the original 48 Squadron pilots. Already the holder of the Military Cross for his recent work, Price and another observer had shot down a two-seater on this morning, and now, with Claye in the back, Price had scored victory No.7. However, Lothar von Richthofen attacked them, wounding both Price and Claye and forcing them to make a crash-landing north-east of Fampoux. Their Bristol, A7110, was a write-off, but at least they had come down in Allied lines. Price saw no more active service but Claye did.

Jasta 11 had no apparent losses, although Jasta 4 had Leutnant Hans Klein wounded, in Albatros DIII 794/17. It was the first of three wounds this ace received and it certainly looks as if 48 got him this day. Klein came from Stettin and was 26; three days earlier he had scored his ninth victory. He would end the war with 22 victories and the Pour le Mérite.

Also on the ninth, Wilhelm Cymera was killed. This 31-year-old Westphalian had brought down Quested and Dicksee, it will be recalled, back on 27 December. Now, with a possible five victories whilst flying with Jasta 1 he was killed in combat over Chamouille, down on the French front. He may have been downed by Adjutant Lucein Joseph Jailler of N15. Jailler claimed a German aircraft between Chamouille and Bruyères, in flames, at 18.15 for his ninth victory. He would achieve 12 victories and receive the Medaille Militaire and Légion d'Honneur.

As if to even the score, the French lost an ace of a tri-place machine (three-seater) on the 10th. Capitaine Didier Lecour Grandmaison piloted Caudron R4s of Escadrille C46 and with his two gunners had scored five victories since mid-1915. The R4 machine was often used as a sort of escort gun-ship. This day, however, they were shot down in combat. Grandmaison and one gunner, Caporal Crozet, were killed, the other gunner, Sergent Boye, slightly wounded. They were downed by Heinrich Gontermann of Jasta 15, better known for his destruction of balloons. The Caudron was his 20th victory, which fell at Berry-au-Bac at 18.20 (G) on a sortie to Juvincourt. Gontermann would achieve 39 kills (17 being balloons) before his death in October 1917, flying one of the new Fokker Triplanes.

Jasta 11 again
May 11 saw Richthofen's squadron again fighting 48 Squadron's Bristols, but this time it was Leutnant Willi Allmenröder (brother of Karl) as well as Lothar who scored. No.48 lost three Bristols this day, two in the afternoon patrol after running into Jasta 11, although the midday loss was also to Jasta 11.

In fact the first casualty was the rising pilot John Herbert Towne Letts, from Lincoln, who had gone through Sandhurst and the Lincolnshire

Regiment before becoming a pilot. As a Brisfit man he would score 13 victories, only to die in a crash flying an SE5a in late 1918. At 12.45, Letts and his observer, Second Lieutenant J Jameson shared an Albatros Scout with another crew over Biache, but were then hit and came down inside Allied lines near Willerval. Leutnant Otto Maashoff of Jasta 11 claimed a Bristol down at this location, his first of three kills. Letts was not an ace yet, but he soon would be, having survived this encounter.

The afternoon sortie took to the sky at 15.00, heading for Arras. They ran into Jasta 11 in the Izel-les-Equerchin area, west of Douai. Although Lieutenant Holliday and Captain A H W Wall claimed one Albatros crashed and another out of control, Allmenröder and the younger Richthofen both shot down Bristols. Richthofen's adversary crashed at Izel at 17.10 German time, Willi Allmenröder's coming down near Beaumont five minutes later. The British crews lost were Lieutenant William Otto Braasch Winkler, another 48 original from Edinburgh, and observer Second Lieutenant E S Moore from Hull. Winkler was killed but Moore survived as a prisoner. This team had scored six victories. The other crew, Captain Arthur Tulloch Cull and 1AM A Trusson, were killed; both Londoners and both married. It seems likely that Richthofen brought down Cull, Allmenröder Bill Winkler. It was Lothar's 23rd victory, Allmenröder's second, and last, for he was wounded on 24 May. British AA saw one Bristol fall in flames near Fresnes, and another go down near Gavrelle apparently under control. The latter was probably Winkler, mortally wounded.

The Germans lost an ace too on this 11 May. Offizierstellvertreter Edmund Nathanael of Jasta 5, victor in 15 combats, was shot down in a fight with 23 Squadron's Spads between Baralle and Bourlon at 20.15 German time. His victor was Captain W J C K Cochrane-Patrick MC flying B1580, his claim being timed at 19.05, west of Bourlon Wood. The Albatros went down in flames, shedding its wings as it did so. It was this Irish ace's ninth of an eventual 21 victories (which would bring him the DSO and a Bar to his MC), who was just short of his 21st birthday. Nathanael was 27, from Dielsdorf, and had served with FFA42 and Jasta 22 before joining Jasta 5. His 14th victory had been the first SE5 to be brought down by a German pilot, on 30 April 1917.

On 13 May Lothar von Richthofen was knocked out of the air war for five months. After downing his 23rd victory, he was hit by AA fire and wounded in the left hip, but put his machine down without further mishap. The 13th was to prove an unlucky date for Lothar in this war.

MAY – JUNE 1917

New aircraft

The Royal Flying Corps had fought Bloody April largely with the same aircraft they had employed the previous summer, which will be remembered mainly for the Somme battles that commenced on 1 July 1916.

The fighter pilots had won the fight against the Fokkers already, with the FE2b and FE2d, the DH2, the Nieuport Scout and the Sopwith Pup. Of these the DH2 was now out dated and would soon be absent from the Western Front. The Pup too was fast becoming less of a match for the German Albatros Scouts, but some units would struggle on with them for some months to come. The new SE5, to be followed shortly by the SE5a, was one of the new fighters the RFC were now using, and the Camel would be along in the summer. The Naval fighter squadrons attached to the RFC still used the Pup and also the Sopwith Triplane. During mid-1917 the Triplane would do fairly well against the Germans, although the RFC never used it. They continued to use the French Nieuports, the types XVII and XXIII, and another French fighter being used extensively now by the French themselves, the Spad VII.

The last DH2 ace to fall

For some reason the DH2 was still operating as late as May 1917 causing problems for its pilots. While its 'big brother' the two-seat FE continued in action and to some good effect, the DH2 was long past its prime. It had fully equipped 24, 29 and 32 Squadrons, but while 24 were changing to DH5s and 29 had moved to Nieuports, 32 were still operating with DH2s whilst changing to DH5s. Not that the DH5 would be terribly good, and it only lasted till the autumn.

Perhaps one of the last DH2 fights cost 32 Squadron one of its original pilots, Captain William George Sellar Curphey MC & Bar. His modest score of six belies his aggressive attitude to air fighting, but his decorations don't. A few months earlier and he would have been given a DSO. With 32 Squadron being based at Léalvillers, between Amiens and Bapaume, and Jasta 2's (Boelcke) airfield at Pronville, just east of Bapaume, it is not surprising 32 and Jasta 2 met on frequent occasions. One such occurrence was on 4 February 1917, the two groups meeting near Achiet-le-Grand, just north-west of Bapaume. 'Growler' Curphey drove down one enemy fighter out of control, but several of 32's machines were hit. The patrol disengaged and returned to base, where Curphey and Second Lieutenant A C Randall (future ace) changed machines and, with others, went back in search of the hostiles. Whether it was the same bunch or not, this time they had several combats with Jasta 2 pilots.

One Albatros was shot down by Curphey and two other pilots, which was probably Leutnant Franz Christian von Scheele of Jasta Boelcke, who was killed, but Curphey was then hit in the head by fire from Leutnant Erwin Böhme over Le Transloy, south of Bapaume. Despite his wound, Curphey got down on the right side of the line. Böhme claimed his tenth victory (he added another before the day was out) but Curphey was not seriously hurt and survived to fight another day.

Now, on 14 May, flying DH2 A2622, he led a low patrol to attack a balloon east of Cagnicourt, just five kilometres north of Pronville. The de Havillands were engaged by Jasta 2 as Curphey and Lieutenants StC C Taylor (a future ace) and Wright headed for the German balloon lines. The observers in the balloons were seen to throw out their maps and papers, then descend by parachute as the balloons were hauled down. Six Albatros Scouts fell on them, and although Taylor claimed one destroyed, Curphey was seen to be hit and go down. Just 20 feet from the ground his DH2 erupted in flames, hit the ground and turned over. It was 11.15 German time, and Hauptmann Franz Josef Walz, the Staffelführer, had achieved his seventh and final victory, north of Séverin Farm.

Walz, from the Mannheim area, was 31, and had been with a Bavarian infantry regiment since 1905, being commissioned three years later. He learnt to fly pre-war, and when war came he served with FFA3 and later Kasta 2 where he gained six victories. Taking command of Jasta 2 on 29 November, this was his only victory with this unit, and later he moved to Palestine, to command FA304. In 1918 he received the Pour le Mérite, having made more than 500 war flights. He died after WWII as a Russian prisoner of that war.

'Père' Dorme falls

A number of French airmen became legends in their own lifetime, and so did some French units. For instance the famed Cigognes Group – the Storks – was very well known by 1917. Among its pilots its four Escadrilles (Squadrons) Spa3, Spa26, Spa73 and Spa103, could count the most famous, names such as Georges Guynemer, Alfred Heurtaux, Alfred Auger, Albert Deullin as well as the French 'ace of aces' René Fonck.

In the early days one man stood head and shoulders above the rest despite his mere 22 years on this earth. René Pierre Marie Dorme, known affectionately as 'Père' (father), came from the Verdun region. He entered military service in 1913; assigned to the 7th Artillery Regiment he saw service in North Africa and by 1914 held the rank of Maréchal-des-Logis. Requesting a move to aviation he became a pilot in May 1915, his first unit being C94. His dedication and aggressive actions brought him to the attention of Felix Brocard who commanded the Storks and he was requested to join his élite group in June, going to N3, flying the nimble Nieuports. His action brought him the Medaille Militaire, the Légion d'Honneur and Croix de Guerre avec Palmes, as well as a commission.

With a score of 23 official and at least 17 unconfirmed victories, he made his last flight on the evening of 23 May 1917 (take-off 18.40) flying a Spad VII, with his companion Albert Louis Deullin (20 victories by 1918). They were in a fight after which Deullin could not see Dorme, but he later spotted a burning Spad on the ground. Neither he nor the rest of the Groupe believe Dorme could have been overcome in the air. He had been in more than 120 combats, while his flying hours exceeded 620, but he had indeed fallen.

They had been in a fight with pilots of Jasta 9, in which a young German pilot of almost identical age had achieved his fifth victory by downing a Spad VII at Fort de la Pompelle, on the German side of the lines. The pilot's body was later identified as Dorme's. The German was Heinrich Claudius Kroll, from Flensburg, near Kiel, the son of a schoolmaster. He had transferred to aviation from the infantry and been commissioned in 1915. After leaving Jasta 9, Kroll commanded Jasta 24 in July 1917 until a wound took him out of the war in August 1918. He had achieved 33 victories and won the Pour le Mérite.

A close call; or 1 Squadron nil, Germans two

June 1917 began with two future British aces almost falling on the 3rd. No.1 Squadron flew the Nieuport Scouts and amongst its embryo fighter pilots were Lieutenant Louis Fleming Jenkin and Leslie Morton Mansbridge, the former from London, the latter from Wales.

Just as the British had its Royal Naval Air Service, so too the Germans had its Naval Air Service and Marine Corps. While in the main the former operated over the North Sea and even made raids upon England, the Marine Jagdflieger units operated along the North Sea coast of Belgium and inland over Belgium and northern France. Marine-Feldjagdstaffel Nr.I had been formed in February 1917 under the command of Leutnant zur See Gotthard Sachsenberg, a future Pour le Mérite winner and victor in 31 air fights. Based at Aertrycke, it flew Albatros DIII fighters.

On 3 June, 1 Squadron got into two fights, the first with aircraft of MFJI just north of Ypres. 1 Squadron's patrol began at 16.00, Jenkin flying with three other pilots and over Zandvoorde two German fighters dived on them. Jenkin's machine (B1690) was shot through the fuel tank. He had to make a forced-landing, coming down safely, but overturned on landing, damaging the Nieuport, although he was unhurt. Flugmaat Bertram Heinrich claimed a Nieuport down inside British lines, his fourth of an eventual 12 victories.

The British Squadron's other patrol left at 1845, along the same patrol line, Langemarck-Becelaere-Tenbrielen, Mansbridge flying B1639, in company with two others. Over Gheluwe at 12,000 feet, five German fighters dived on them and a combat ensued, but Mansbridge was wounded in the thigh and his machine badly shot about, but he got home.

Again it is thought that MFJI were the opposing force, Flugmaat Kunstler claiming a Nieuport north of Ypres, which was probably Mansbridge, although all the Nieuports managed to get back across the lines. It was the German's first claim. However, had not fortune looked favourably on 1 Squadron this day, neither British pilot would have become more than a war statistic. As it was, Jenkin went on to score 22 victories, Les Mansbridge returning to 23 Squadron in 1918 to down his fifth.

The FEs' summer
The FE2s were still battling it out with the Germans on an almost daily basis and, despite their antiquated appearance, were still a match for the most ardent of German jagdflieger. By the early summer of 1917, although nearing a date for re-equipping with Bristol Fighters, two units still fought on with their FEs – Nos. 20 and 22 Squadrons. And as they were to show in June and July, they could still hand out a nasty shock to the opposition. On 5 June, two German aces went down to FEs, Leutnant Karl-Emil Schäfer, leader of Jasta 28, and Leutnant Kurt Schneider of Jasta 5.

Schneider came from the Leipzig area and was 28. He had been decorated with the army before joining the air service, and was one of the original Jasta 5 pilots in August 1916, although he did not score his first confirmed victory until March the following year. But then he scored consistently until his tally reached 15 by the end of May. On an early morning patrol by 22 Squadron led by Captain Carleton Main 'Steve' Clement and his observer, Second Lieutenant L G Davies (in A5461), they met Jasta 5 north-west of Lesdain, south of Cambrai, at 07.15. In the fight, Clement and Davies claimed two Albatros Scouts shot down, one seen to crash, the other out of control. In one of them was Schneider, mortally wounded. He died later that same day.

Clement, a Canadian, was the commander of C Flight, and by this time the longest serving surviving member of the Squadron. He was just 21 when he met Schneider, but had already served in the Canadian army and been flying FEs since the summer of 1916. He won the MC and French Croix de Guerre, but his fight with Jasta 5 was his last on FEs, 22 Squadron going over to Bristol Fighters in July. By August he and his observers had accounted for 14 German aircraft, but he was killed by AA fire from Leutnant Böhner's Flakzug 99 (in A7172) on the evening of 19 August, over Langemarck.

MAY – JUNE 1917

Krefeld-born Schäfer, a former Jäger Regiment NCO, had made his name with Richthofen's Jasta 11 after flying with Kasta 8 of KG3. By the end of Bloody April, this 25-year-old had scored 23 victories and been given command of Jasta 28, where he had raised this to 30 by 4 June. No doubt he was looking for victory No.31 the next day but ran into 20 Squadron.

This Squadron had sent out two small formations of three and four FEs shortly before 14.00, although one of the four had to abort with engine trouble. Leading this first section was Captain Harold Leslie Satchell, who had previously been with the Royal Warwickshire Regiment. He had seen action in April and by early June had scored just three victories (of an eventual eight). His usual observer/gunner was Lieutenant A N Jenks, but on this occasion, flying FE2d A6469, he had Lieutenant T A M S Lewis in the front cockpit, who would achieve six victories. In the 1930s he would be known as T A M Stuart Lewis MC.

The FEs patrolled the Warneton area, taking the time to take 11 photographic plates of items of interest on the ground. Visibility was good so there was no surprise attack as 15 German fighters were seen approaching. Leading them was a red machine, so quite obviously Karl Schäfer had taken a leaf out of his old commander's book and led his new Jasta in a red-marked machine, or perhaps he had even taken his old Jasta 11 Albatros with him. However, at 14.00 the previous day (the 4th) Schäfer had flown alone to test the range of his guns and had seen three Sopwith Triplanes and engaged them. One of his elevator cables was shot through but he made a good landing near the front lines. He had previously been shot down back on 28 May '. . . in his other red machine . . .' very near to the location where he was about to engage 20 Squadron.

According to the Germans after the war, Schäfer was some way ahead of his pilots when he dived on the FE's, taking them on alone at first, although 20 Squadron make no reference to this. Schäfer attacked the FE flown by Lieutenants William Wright Sawden RGA/RFC and R McK Madill (A6384), mortally wounding the 26-year-old pilot and forcing him to dive for the lines. (He landed near Ypres without further injury but succumbed to his wounds.) The red Albatros began to follow – the classic fatal mistake. Satchell at once dived to the rescue, Lewis firing into the Scout, whereupon the German broke off his pursuit and began to dog fight the FE. The fight continued for some 15 minutes, Satchell later recording that the German pilot showed great skill and persistence. Following a burst of fire from very close range the Albatros burst into flames and all four wings were seen to rip off before it crashed near Becelaere, east of Ypres. It was 16.05 German time (15.05 British).

Post-war the Germans confirmed that the time was 16.05 (G) and that other witnesses saw the right wing break away and the Albatros crash headlong vertically into the ground. Of the top right wing only the aileron was found, the left wing falling a long way from the main crash site, noted as being Zandvoorte (south-east of Ypres). That night the body was recovered and taken to Lille. Every bone in the German's body had been broken, the skull smashed to pieces and the heart ruptured. There were no visible bullet wounds. Therefore Schäfer, like so many other WWI fliers, had to endure his final moments on earth, uninjured, but knowing he was going to die. At least he was spared the torment of fire as he plunged earthwards.

The third German ace to fall to FEs was Leutnant Ernst Weissner, a 22-year-old from Stuttgart, serving with Jasta 18. On 7 June, the day the Battle of Messines opened, heralded by 19 mines being exploded beneath the German lines (through excavations dug from the British trenches), he celebrated his seventh victory, a 6 Squadron RE8 at 17.10, but it was short-lived.

That same evening, Weissner was either out again or still out, and found another British two-seater over the front, but when about to attack he was seen by a patrol of 20 Squadron. Captains Frederick James Thayre, from London, and Francis R Cubbon were, like Clement of 22, the most experienced men on the Squadron. Thayre had started as a Corps pilot with 16 Squadron but had then moved to FEs and 20 Squadron. By this date, they had accounted for 20 hostile aircraft, all but two being single-seat Albatros Scouts. Seeing this latest Albatros heading for the two-seater, during their evening patrol over Houthem (in A6430), they dived, Cubbon firing his Lewis gun into the Albatros, and it nose-dived. Just like Schäfer's machine two days earlier, it shed its wings before crashing into the ground – victory No.21. The Germans timed Weissner's death at 17.50, coming down near Wambake between Warneton and Comines.

Navy versus Marine

The RNAS lost a promising pilot on 8 June, Flight Sub Lieutenant Thomas Gray Culling, a 21-year-old New Zealander (just eight days past his 21st birthday) who had achieved six victories with 1 Naval Squadron, flying Triplanes, often in company with his flight commander, Captain R S Dallas DSC. Culling too had just received the DSC but flying on patrol near Warneton (N5491) he was shot down by Vizeflugmeister Bossler of MFJI and killed. It was the Marine pilot's first victory, timed at 11.25 (German), and apparently his only one with this unit, for he was wounded on 14 June. Later he moved to MFJII, scoring two more kills before the war ended.

MAY – JUNE 1917

The Sopwith Pup units were still going strong, although they would soon start to be replaced with the new Sopwith Camel. 66 Squadron had been battling away with Pups for some months, and one of its pilots, London-born Captain James Douglas Latta MC, had claimed five victories with his two previous Squadrons in 1916, 1 and 60. He became a flight commander with 66 in May 1917 but was wounded on 8 June. Flying Pup B1726 on an OP to Roulers, they got into a scrap with Jasta 8, and Latta was hit and wounded by Oberleutnant Bruno von Voigt, although he misidentified the Pup as a Nieuport. Jasta 8 claimed two other aircraft in this action, both correctly as Pups, Oberleutnant Konrad Mettlich reporting two Pups destroyed. These were undoubtedly A6207 and B1745 which collided during the fight. Another Pup was badly shot up but its pilot got back. These were Mettlich's first two kills of an eventual eight, but as we shall read later, he did not survive the war.

Thayre and Cubbon go down
The big pusher FE2 machines had been in action over France for a long time and, while dated by that time, could still give a good account of themselves in experienced hands. They would soon be replaced by the Bristol Fighter which had already been in France since April 1917, although inexperience with how to handle them in combat had caused losses. However, the FE pilots and gunners were still a force to be reckoned with. A number of crews had built up creditable scores with their FE two-seaters, one being the team of Captain Frederick James Harry Thayre MC and Bar and his usual observer/gunner, Captain Francis Cubbon MC and Bar. Thayre, a 23-year-old Londoner, had achieved 20 victories flying with 16 and 20 Squadrons, 19 with the latter.

As well as air fighting and reconnaissance, the FEs also operated as day bombers (later night bombers as well), and on 8 June 1917, six of the Squadron's FEs lumbered off the ground at 05.30 hrs to bomb Comines while flying a Central OP, led by Thayre. Not long after they had crossed the lines east of Ploegsteert, a two-seater Albatros was observed and Thayre dived on it. It was last seen going down in a vertical nose-dive with smoke streaming out but was not seen to crash. However, as the FE (A6430) began to recover and gain height, it was hit by anti-aircraft fire and fell to the ground near Warneton. The FE was claimed by a German AA unit – KFlak 60 – both men being killed.

The Staffelführer gets his third
Oberleutnant Kurt-Bertram von Döring, aged 27, had joined the Dragoons in 1907, but transferred to the Air Service in 1913. In 1914 he was flying with FA38, then served with two other units prior to becoming

a fighter pilot. Just as the Arras Battle was starting he took command of Jasta 4 despite not having scored any victories by this date. By early June he had rectified this by having two kills. Then on 9 June 1917 he gained his third by downing an ace of No.1 Squadron RFC.

Eight Nieuport Scouts headed out on a Northern OP, covering the area Houthem–Menin–Dadizeele, taking off at 13.27. The patrol comprised a number of aces or coming aces – Captain W T Campbell (leader), Lieutenants W T V Rooper, W W Rogers, L F Jenkin, and F Sharpe, Sergeant G P Olley, plus two more officers. Over the lines they were to have fights with at least three formations of Albatros Scouts, one of six, another of seven then one of four. At about 15.00, Frank Sharpe (B3481) dived on a formation of six enemy scouts, followed by Campbell, and attacked a black and white Albatros. Campbell fired into it and it fell away in a spin and crashed.

Lieutenant R W L Anderson also became involved, tagging on to Sharpe, as the enemy scouts scattered but then more Albatroses arrived from all directions, until around 17 were in the fight. Anderson saw a Nieuport get below most of the Albatros machines and dived to try and distract them, but the Nieuport was hit and went down in a spin with three Germans on its tail. Anderson tried to intervene but was unable to get his sights on any of the Germans, and his last sight of Sharpe was of him zooming over trees and houses. Anderson then had to change the drum on his Lewis and then succeeded in chasing the three Albatros Scouts off – more probably they were low of fuel.

Jenkin had been with Sharpe too, east of Houthem, and shot down one Albatros, seen to crash by Gordon Olley. Olley also forced one to land near Houthem, and another black Albatros became a flamer over Dadizeele. Sharpe failed to return, last seen near Gheluwe, and later, 41 Squadron pilots reported seeing a Nieuport land under control. However, he had got down safely and became a prisoner. Just two days earlier the Londoner had destroyed an Albatros and flamed a balloon to raise his score to a modest five. Von Döring reported his victim down at Zandvoorde, so Sharpe had obviously been heading for the lines when he was forced to make a landing. He would go on to achieve 11 victories and become a Rittmeister, occasionally leading JGI when von Richthofen was on leave. He also commanded Jagdgruppe Nr.4, Jastas 66 and 1, surviving the war. Post-war he was an instructor with the Argentinian and Peruvian Air Forces and later with the German Mission to China. With the Luftwaffe he rose to Lieutenant-General in WWII.

Upon his return from Germany in December 1918, Sharpe stated: 'I dived to attack an EA which was hit and crashed. Then for reasons unknown I crashed myself. Recovered consciousness in hospital four days later with no recollection of what had happened.'

Willie get his fifth
Hermann Becker was only a few weeks short of his 30th birthday but he was a budding fighter pilot with Jasta 12. Although he would end the war with 23 victories and be nominated for, but not receive, the Pour le Mérite, he was lucky to survive an encounter with 60 Squadron on 16 June 1917, his score being just a modest one.

A patrol of Nieuport Scouts of 60 Squadron had taken off for the evening patrol to Marquin, Keith Caldwell, Willie Fry, Jack Rutherford, D C R Lloyd, J Collier and W E Young being among the pilots. At 20.30 hrs above Marquin the patrol spotted two Albatros Scouts and dived, chasing the two single-seaters down to 3,000 feet, several pilots firing intermittent bursts. Willie Fry, a 20-year-old with just four victories to date, lined up and fired 40 rounds from his Lewis gun at one of the Germans which had eased up and slid in behind Lloyd as he overshot. Jack Rutherford also opened up at close range at one Scout, then fired on the one attacking Lloyd. Both Lloyd and the second German disappeared in the evening gloom, and Lloyd failed to return. The British pilots claimed one of the Germans 'out of control', shared by Fry, Caldwell and Collier, but Willie Fry now had his fifth victory; he would end the war with 11 and the MC. Caldwell would score 25, and command 74 Squadron in 1918.

Becker was seriously wounded but got down. He would return to Jasta 12 in the summer. He was, however, an experienced airman already, having flown as an observer before becoming a two-seat pilot. He had only been with Jasta 12 a month when wounded. Lloyd was killed. He had in fact been in collision with the other Albatros, flown by Vizefeldwebel Robert Riessinger, who also died.

CHAPTER FOUR

June – July 1917

Black Flight lose a pilot
For some time now, the Royal Naval Air Service had reinforced their brothers in the RFC with fighter squadrons on the Western Front. They had initially flown Sopwith Pups and Nieuport Scouts but by mid-1917 were flying Sopwith Triplanes. This was a pretty good machine for the period, and it is amazing that the RFC did not decide to use it. Nevertheless, the RNAS boys did extremely well, none more so than 10 Naval Squadron.

Ten Naval were commanded by Squadron Commander B C Bell DSC and had re-equipped with Triplanes in May. Its 'B' Flight was commanded by Flight Lieutenant Raymond Collishaw, a Canadian, who would end the war as one of the Allied top aces. In due course Collishaw's men gave their machines various names starting with Black – Black Maria, Black Roger, Black Death, Black Prince, Black Sheep and so on. The front part of the Triplanes was also painted black.

Collishaw's was Black Maria, not a bad name considering the number of German airmen he despatched, and Gerald Nash's machine was Black Sheep. Nash was also a Canadian and had claimed seven victories in May and June 1917, but he did not see out the month.

On 25 June Collishaw (N5492) led his six men on a patrol, taking off at 17.30 (Collishaw, Ellis V Reid, Nash, Ray Kent, Wm M Alexander and Des Fitzgibbon). The weather was not good, so acting on instructions, Collishaw did not take his pilots over the line but flew a line patrol instead, at 7-8,000 feet between Armentières and the flooded area around Dixmude. Three red German scouts were spotted near Hollebeke and an indecisive combat took place. Flight Sub Lieutenant Kent (N5357) lost contact with the formation and returned early. When the Naval boys re-formed Gerry Nash was not about. The German pilots were from von Richthofen's Jasta 11 and Karl Allmenröder got in some telling shots which crippled Nash's Triplane (N5376), forcing him down east of Messines. He had last been seen by Allied observers landing east of Comines at 18.10, under control after combat with enemy aircraft. It was Allmenröder's 29th victory, reported down west of Quesnoy at 18.45. Fortunately Nash was unharmed but spent the rest of the war a prisoner.

Allmenröder was himself shot down and killed on 27 June, hit, it is assumed, by AA fire. He fell into no-man's-land near Zillebeke, German soldiers bringing in his body when darkness fell. His fall occurred at 09.45 and some years later Ray Collishaw suggested in his book *Air Command* that he had fired the fatal shots which brought him down. He said it was a long-range shot and he did not claim a victory, but it is now accepted that the German ace fell to ground fire. Collishaw's patrol headed out at 17.20, and he fired at 'his' Albatros some 55 minutes later. By this time, Allmenröder's body had been lying in his smashed fighter for around eight hours.

The FEs of 20 Squadron
Considering the big two-seat pusher FE2s had been around some time, they could still hold their own against German fighters. Indeed, some of the more aggressive FE squadrons, such as 20, 22 and 11, were no pushovers, especially if they held formation or went into one of their defensive circles. The gunner/observer in the front cockpit of the gondola had a terrific forward field of fire with no propeller to block his view, and the pilot too had a fixed Lewis gun firing forward. And if attacked from behind, the observer could stand on his seat and operate a rear-firing Lewis gun mounted on a metal rod which could fire back over the wings and engine.

In the summer of 1917 these pusher crews would account for a number of German aces. One of the first was Walter Göttsch of Jasta 8. Göttsch, it will be recalled, had already been wounded by 20 Squadron back in February. Having returned to active duty in April, the German had now raised his score to 12. He seemed attracted to 20 Squadron's FEs, for since his return to combat he had downed three, making five in all since January. Now they were about to get him again and help even up the score.

On 29 June 20 Squadron flew a Central OP comprising six FEs, taking off at 11.26. The FEs always looked for trouble and almost always carried bombs and took a camera-equipped machine. However, two FEs dropped out early with engine trouble leaving four to complete the assignment. In the vicinity of Houthulst, they were attacked by Jasta 8, after dropping bombs and taking ten photographs. As the Albatros Scouts roared in they were met by a wall of fire from the FEs which immediately knocked one fighter out, which went spinning down on fire. A second Scout was also reported down on fire, shedding its wings near the ground, but this was probably the same machine seen from a different angle – a not uncommon visual occurrence in the heat of combat.

The German pilot was Leutnant Alfred Ulmer, who was not going to see his 21st birthday in September. He had five victories but had not scored since January. It will be remembered that he had been severely wounded by 46 Squadron on 5 February. He crashed in flames, timed at 13.15, credited to Lieutenant H W Joslyn and Private F A Potter (A6415). He was pulled from his wrecked Albatros DV, miraculously still alive, but he died of his injuries soon afterwards.

The other German pilots closed in on the FEs only to have a second DV badly hit. This time it was Göttsch. He might have guessed. His machine crippled, the FE crews saw the fighter go down in a spin near Houthem, this one credited to Lieutenants R M Makepeace and M W Waddington (A6498). However, they were credited with an 'out of control' claim, their first victory. Reg Makepeace would go on to score 17 by early 1918. Waddington too would become an ace gunner with 12 victories.

The rest of Jasta 8 became understandably reluctant to press home an attack on the FEs and soon broke off, just firing from long distance but refusing to come in close. In this way the FEs headed home, having downed two German aces, one dead and one shaken up. Göttsch, the latter, got his revenge on 31 July by downing yet another 20 Squadron FE, and ended his war with 20 victories. On 20 April 1918 he took on a less aggressive adversary, or so he thought, an RE8, only to have his Fokker Triplane severely hit. He fell to his death into the British lines near Gontelles.

Another German ace to fall at this time was Albert Dossenbach who had only recently taken command of Jasta 10, within the framework of the newly formed Jagdgeschwader Nr.1, led by Manfred von Richthofen. Dossenbach was 26 and came from the Black Forest region. He had achieved 14 victories as a two-seater pilot, then a fighter pilot with Jasta 36. He flamed a balloon on 27 June, his first with Jasta 10 and his 15th overall, but met his fate on 3 July.

Four DH4 bombers of 57 Squadron had flown a sortie, taking off at 10.45, led by Captain L Minot and his gunner, Lieutenant A F Britton (A7487). Over Zonnebeke at 12.25, now down to three 'Fours', one having dropped out with engine trouble, they were engaged by four Albatros Scouts. As they came in, one Albatros was seen to turn over and spin away but then Minot was attacked by another from underneath. Minot banked his machine to allow Britton to fire down at their antagonist, his bullets setting the German machine on fire. The Albatros began to tumble earthwards, Dossenbach then leaping from the burning mass rather than face a slow cruel death by burning. His body was recovered and returned to his native Germany for burial.

Laurence Minot became a minor ace with 57 Squadron, but was himself killed in action on 28 July, flying with Lieutenant S J Leete (A7540), as Britton had been wounded the previous day. They were last seen diving after enemy aircraft east of Courtrai during a bomb raid on Ingelmunster aerodrome at 18.00 hrs. Three of the seven DH4s failed to return from this sortie, following a fight with Jasta 6, led by Oberleutnant Eduard Dostler. Dostler claimed one DH4, Leutnants Robert Tuxen and Walter Stock the others.

The Ringmaster falls
The FEs of 20 Squadron claimed another success on 6 July, this time the new leader of JGI, Manfred von Richthofen. Six FEs flew out at 09.50 on a Central OP, and took bombs to drop on the Houthem dumps. The six crews estimated a force of some 30 enemy fighters engaged them, as they struggled to return to the lines. So intent were the Albatros pilots that they pressed their attacks closer than usual, probably because it was JGI and all wanted to impress their leader. However, five Albatros machines were claimed shot down 'out of control', meaning (usually) they spun away to get out of trouble. Four of these went down to Captain D C Cunnell and his gunner, Second Lieutenant A E Woodbridge (A6512), the fifth claimed by Second Lieutenants C R Richards and A E Wear (A6498). Three of the FEs received damage and one observer was killed, but all got back.

The odds are that Cunnell and Woodbridge scored the hits on von Richthofen's machine, but it could just as easily have been Richards and Wear. Whatever happened in fact, von Richthofen was hit a glancing blow to the head, which temporarily blinded and paralysed him. Only too aware of his problems, Richthofen was fortunate his Scout did not whip into a spin and crash before his sight returned and he regained the use of his arms. In any event he managed a creditable forced landing but with a serious head injury he was away from the Front for some weeks. He returned to duty at the end of the month but was never really the same man again, having undoubtedly realised that he was not himself invulnerable. It had been a close call, and another inch lower down and 20 Squadron's bullet would have ended his career with a score of 57.

Donald Charles Cunnell was killed in action just five days later, the 12th, during a patrol to Menin/Wervicq. They had taken off at 15.35 – four FEs – but their machine (B1863) was hit by AA fire and Cunnell was killed instantly. His observer, Lieutenant A G Bill, took over the controls and made a very creditable crash-landing. He was understandably badly shaken, but the FE was a write-off. Cunnell and his gunner/observers had shot down nine enemy fighters since May.

The day after the Richthofen fight, in a month that was to have heavy losses in ace pilots, another Naval Triplane was shot down, this one from 1 Naval Squadron, flown by one of the unit's flight commanders, Cyril Askew Eyre. Although born in Britain 21 years earlier, he had joined the RNAS in Toronto, Canada. Initially he flew Pups with 'A' Squadron which became 1 Naval and re-equipped with Triplanes. On 3 July he claimed his sixth victory.

On 7 July 1 Naval were hotly engaged by pilots from Jasta 4 and Jasta 11 and lost three – Eyre (N6291), Flight Sub Lieutenants K H Millward (N6309) and D W Ramsey (N5480) – all killed; Kurt Wolff and Alfred Niederhoff of Jasta 11, and Leutnant Richard Kruger of Jasta 4 were the victors. The fight occurred in the Comines area at around 10.00 British time (11.00 German time), all three Triplanes falling in the area. All were downed within ten minutes, 10.00 to 10.10, but the problem is that a fourth German pilot, Friedrich Altmeier of Jasta 24, also claimed a Triplane in the same area at 10.05 (11.05 German time). It would seem that Wolff shot down Millward over Comines at 10.00, then Kruger hit Ramsey at 10.05, west of Wervicq, Eyre fighting on until overcome by Niederhoff at 10.10. Millward fell at Comines, Ramsey and Eyre at Bousbecque. Altmeier's claim also crashed at Bousbecque.

Clearly there were only three Triplanes shot down but four claims appear to have been verified. While this is unusual, the Germans generally being very keen to verify all claims and give credit only when certain of the victors, this is one of those occasions when the system seems to have failed. There were no shared victories in the German Air Service, and if there were ever two or more pilots claiming a single victory, arbitration usually 'assigned' the victory to just one pilot, the others gaining no credit at all.

This, however, was Altmeier's second kill; he would go on to make 21 by the war's end. Wolff's victory was his 33rd, and last. For Kruger, it was his first and only; he would be killed ten days later. Niederhoff, however, who is generally thought to have downed Eyre, had raised his score to four on this day and would go on to gain three more by mid-July. We will read more of Alfred Niederhoff in this chapter.

A fast scorer stopped

No.56 Squadron had been the first unit to bring the new SE5 to the Front in April 1917, with the redoubtable Albert Ball as one of its flight commanders. Three months later, Captain Edric William Broadberry MC had scored eight victories, along with R T C Hoidge the fastest scorers on the Squadron.

On 12 July the Squadron flew an evening OP to Menin–Roulers at 18.00 hrs led by Captain P B Prothero, Broadberry flying at the right rear of a 'Vee' formation in A8918. The SEs were at 14,000 feet and were surprised by an attack from above and behind. The first Broadberry knew of it was when bullets riddled his machine; his instrument panel shattered, there was a sharp pain in his right calf, and the engine began to splutter and die. Instinctively looking over his shoulder he saw a number of German scouts coming down from the evening sun's glare. He knew he was out of the fight and switched off the petrol supply to his now dead motor and dived for the lines. He managed to get down on Bailleul aerodrome and only later did he discover his leg wound. Two other SE5s had been hit in the attack, J S Turnbull (A4861) and Captain E D Messervy (A8929). Turnbull had been wounded in both legs but force-landed inside British lines; Messervy, hit in the engine, crash-landed at Poperinghe and turned over due to running into a wire fence.

They had been extremely lucky. It seems that Jasta 6, part of JGI, claimed all three, but oddly, the times for each vary; 16.35, 19.40 and 21.45 (German time). The three victories were credited to Hans Adam, his fourth, Kurt Kuppers, his fourth, and the Staffelführer, Eduard von Dostler, his 16th. All three were acknowledged as down on the British side of the lines, but as we can see, Broadberry got back to a British airfield. Each was said to be a Sopwith, but the Germans often confused Allied types, especially if they didn't have the wrecks on their side of the trenches.

To confuse the issue further, Theo Osterkamp of Marine Feld-Jasta I also claimed an SE5, at 21.00 (German time) over Zandvoorde. Jasta 6 said their three came down east of Dickebusch, Wytschaete and Zillebeke. Broadberry later said they were attacked when south-east of Ypres, near Hollebeke. All four therefore are in the general area. There are no other candidates from the British side, other than a Sopwith 1½ Strutter of 43 Squadron, which force-landed after an evening Line Patrol. It had been damaged by an enemy aircraft and the observer was wounded in the leg, but they too had come down inside British lines; probably the 'Sopwith' claimed by Vizefeldwebel Otto Marquardt of Jasta 4 over Zandvoorde at 21.05 German time – that went down on the British side too. Whatever the outcome, Edric Broadberry was out of the war and he did not see further combat. He remained in the RAF, however, and became a Group Captain in WWII.

Hans Adam gets another
As the reader will be realising by now, it is not always easy to tie up who was fighting and claiming whom over the Western Front in WWI. It

should be easier to do so for the Germans, however, for their stricter rules of confirmation, coupled with the fact that most Allied losses came down on their side of the lines, provides better information; their one-victory-one-claimant policy also helps. Nevertheless, the ideal isn't always achieved and while generally the loss and claims tally fairly well, as we have just seen, on some occasions there is considerable doubt due to just too many credits for too few losses, which in theory should not occur.

However, whether or not Hans Adam got one of the 56 Squadron SE5s on the evening of 12 July, he most certainly got a Nieuport on the 13th, thereby achieving his fifth kill. The Nieuport Scouts of 29 Squadron flew an escort mission at mid-morning of the 13th – Friday the 13th – for the Sopwith 1½ Strutters of 70 Squadron. The 'Strutters' were flying a Photo Op over the line Gheluwe–Becelaere–Zonnebeke–St Julian, leaving their base at 09.00. Lieutenant H M Ferreira of 29 Squadron saw 12 Albatros Scouts diving down from the west, heading straight for the two-seaters. Ferreira attacked one lining up a Sopwith and drove it off, then attacked another which spun away, thought to be out of control. Ferreira continued to engage with the other Nieuports, but was then himself driven off. While so engaged he noted three aircraft going down in flames. Two of these were Sopwiths, 43 indeed losing two machines in flames near Becelaere. The other was probably Lieutenant Archibald William Buchanan Miller of 29 Squadron. Just the previous evening, this Staffordshire lad – he was only 20 years old – had claimed two Albatros DIIIs out of control to bring his personal score to six.

Jasta 6 claimed three Sopwiths, timed at 11.35 (10.35 British time), Dostler his 17th, Fritz Krebs two, his seventh and eighth (but probably only one), while Dostler had then downed Lieutenant F W Winterbotham (B1577) and Adam had burned Miller's B1506. Karl Dielmann also claimed a Nieuport, but 29 only lost two. Winterbothom survived as a prisoner of war. In WWII he became famous for his work with the Ultra counter-intelligence secret.

The Germans didn't have it all their own way on the 13th. Heinrich Bongartz, who would end the war with 33 kills flying with Jasta 36, had a run-in with DH4 bombers at 16.30 (G), and was hit in the arm by return fire. He force-landed at Warghem/Audenarde. He had scored 11 victories thus far, but would not claim his next until September. He was probably hit by a 55 Squadron 'Four', A7475, flown by Second Lieutenant P G Kirk and his observer, Second Lieutenant G Y Fullalove. They noted their combat at around 16.00, against a mottled-coloured Albatros Scout over Eyne. After dropping their bombs they were engaged by six German scouts coming at them from the front before swinging round behind them. Fullalove fired a whole drum at the nearest machine and saw it go down, leaving a thin smoke trail, but with more scouts around he could not watch it further. Moments later a DH4 went down, closely followed by a German. This was Leutnant Ernst Hess of Jasta 28 who downed one of 55's machines for his third victory (of 17). Percy Kirk, a South African, and George Fullalove from Surrey, were killed exactly one month later, 13 August, the latter jumping from their burning aircraft. They were probably shot down by Jasta 17's Georg Strasser, his fifth of seven victories.

This same day, 13 July, Philip Fullard of No.1 Squadron had a lucky escape. Flying Nieuport B1666 on an OP to the Zandvoorde–Menin–Lille–Armentières areas that evening – a full Squadron patrol of 12 aircraft – they were engaged by a formation of enemy machines. Fullard fought one and sent it down 'out of control' – recorded as his 11th victory – but then he had a problem. He later recorded having engine trouble, but whether this was due to a mechanical defect, or knowingly or unknowingly he'd been hit in the engine, is not clear. However, he was suddenly out of the fight and gliding towards the lines. He scraped over them safely then crunched into a shell hole and turned over. All four wings were damaged, so too were the ailerons, the wheels were smashed and the axel bent, engine housing bent, longerons and fuselage badly strained. The machine was salvaged but had to be sent to No.1 Air Depot for repair, as '. . . it would take more than 36 hours to repair on the Squadron.' Not surprised!

As the machine was salvaged it would not appear on any loss report for the day, and only B3483 was lost in this fight, flown by Second Lieutenant W C Smith, who was killed, last seen going down over Gheluvelt with three fighters on his tail. Credit for him went to Vizefeldwebel Dahm of Jasta 26, timed at 08.05 (G) north-east of Langemarck, his first of four victories. As no other Jasta 26 pilots claimed in this fight, one must assume that Fullard did in fact have mechanical trouble, unless a German claim was not upheld. However, Fullard had survived, and would go on to be one of the big aces of WWI with 40 victories before an unfortunate accident during a football match cost him a broken leg and a return to England. He later rose to Air Commodore CBE DSO MC and Bar AFC.

Another ace to fall on the 13th was Frank Neville Hudson of 54 Squadron. Neville Hudson was a 19-year-old youth from Beckenham, Kent (he wouldn't be 20 until November). Despite his tender years he had already seen service with the East Kent Regiment, then after becoming a pilot he had flown with 15 Squadron, winning the MC as well as being wounded. Moving to fighters he joined 54 Squadron, flying Sopwith Pups, and had scored six victories between January and July. Returning from PoW camp in late 1918, Hudson recalled:

> 'I escorted a recce aircraft over Bruges but my engine cut-out; on turning to glide home, I was attacked by five hostile aircraft and forced down. Whether my engine had been hit by AA fire or not I'm unsure, but I was not under very heavy AA fire at the time. I crashed my aircraft on the beach and it turned over on to its back, but I was unable to set fire to it as I was immediately surrounded by soldiers before I could even get out.'

The five German fighters were from Jasta 20 and, crippled or not, Hudson's Pup was claimed by Leutnant Gerhard Wilhelm Flecken, which he noted as falling at De Haan at 20.30. Hudson had taken off at 18.30 (in A6240) for this Special Recce escort over the Bruges/Ostende area and had last been seen over Bruges and heading for Ostende. He received a Mention in Despatches while in PoW camp, and survived the war.

Flecken had been with Schutzstaffel 15 before joining Jasta 20 in October 1916, Hudson being his second victory. A Camel fell to him in September 1917 but then he took command of Jasta 43 at the beginning of November. With his score raised to four he took command of Jasta 55 on 1 June 1918 in the Middle East. He too survived the war.

No.1 Squadron in the wars again

Two days later (the 15th) 20-year-old Charles S I Lavers of No.1 Squadron, who had a modest score of three victories thus far but would raise this to nine by the war's end, was lucky to survive a combat. In some ways his was not a dissimilar experience to that of fellow pilot Philip Fullard on the 13th. Lavers had been part of a Special Distant OP along a line Polygon Wood–Menin–Lille–Brenchies at 19.20 (B3485). With his five companions he had been embroiled in a dog fight over Menin at 20.10, Lavers having his petrol tank shot through. Luckily there was no fire, and although slightly wounded too he managed to scrape back over the lines to make a forced landing. (This was his second wounding, having been hurt while an observer with 23 Squadron in 1916.) His starboard wing spar had also been shot through, but he survived. He later became a flight commander and won the DFC.

No.56 Squadron shoot down Krebs and Hermann Göring

The very next day Vizefeldwebel Fritz Krebs, an eight-victory ace of Jasta

6, was shot down by 56 Squadron. Krebs was another 20-year-old and had joined Jasta 6 in May 1917, scoring his victories in just three months; he went down north of Zonnebeke at 19.45 (G). A patrol from 56 were escorting Martinsydes to Moorslede by way of an OP. Over Polygon Wood, Captain G H Bowman (nine victories thus far), flying SE5 A8900, after seeing the 'Tinsydes' drop their bombs, had an indecisive fight with four Albatros Scouts.

Some time later, still with three other SEs with him, Gerald Bowman was beneath a patrol of Spads, then saw enemy machines over Roulers. Bowman went for them, but just before diving, checked the sky and found 15 more fighters north of them. Bowman turned his men towards the lines but, not unusually, there was a strong west wind blowing. As they struggled westward, six more German fighters came along from the direction of Menin. The Germans had a 2,000-foot height advantage, not to mention a five-to-one superiority in numbers. Now in a spot, Bowman turned and fired off a red flare to attract the Spads and some nearby Pups. As if a signal for the opposition, the Germans dived towards the SEs, which then adopted dive-and-zoom tactics to avoid close fighting.

There were so many German fighters that it was impossible to try and retain height, and the SEs were forced down. Bowman got behind several Germans but was never able to see the results of his fire as he was constantly chased off. Finally, however, he found himself behind one Albatros and fired both his Lewis and Vickers guns at very close range, whereupon the Albatros nose-dived with engine full on. The pilot made no attempt to flatten out and flew straight into the ground at the east end of the old racecourse by Polygon Wood and smashed to pieces. The SEs then managed a fighting retreat. Bowman would achieve 32 victories in WWI.

In this same fight had been aircraft of Jasta 27, headed by Hermann Göring. Göring (2049/17) had damaged the SE5 flown by Lieutenant R G Jardine (A8931) in a head-on attack. Jardine's oil reservoir was punctured which resulted in his engine bearings seizing. He made a force landing attempt but in pulling his aircraft up to try and skip over a road, he stalled and crunched into a cornfield. Göring claimed him (victory number nine), and indeed the SE5 was struck off charge, but Göring too was in trouble. His Albatros was so heavily damaged in this head-on action that the engine all but shook itself free of its mounting. Göring spun down east of Ypres and his machine was destroyed in the subsequent force-landing which resulted in a somersault, although the future leader of the German Luftwaffe survived unhurt. Robert Jardine was killed in action four days later without ever knowing he had achieved a 'victory'.

On the beach

July was proving an expensive month for aces. On the 17th Lieutenant Roger Bolton Hay MC, a Bristol Fighter pilot serving with 48 Squadron, flying with his observer, Second Lieutenant Oliver J Partington, was brought down by Vizefeldwebel Adolf Werner of Jasta 17. Hay had five victories by this date, following service with the West Yorks Regiment prior to becoming a pilot. The Wiltshire lad and his south London backseater had taken off at 07.05 to carry out a photo-reconnaissance sortie to the Ostend area.

They were intercepted by Jasta 17, Werner forcing them down on the beach at Middlekerke, north-east of Nieuport. Another machine flying not far off saw the action and also saw three men approach the downed 'Brisfit', and five men walked away. The Squadron later thought both men had survived, but in fact Hay had been seriously wounded and died soon afterwards. Partington, who had, like his pilot, been with the Squadron some months, became a prisoner. It was Werner's first of two kills, having been with Jasta 17 for just a week. He remained with his Jasta until May 1918, although both his victories were achieved in July 1917.

The next ace to fall was in an action again on the more northern sectors of the Front. On 20 July, Nieuport Scouts of 29 Squadron flew what was termed an Outer Patrol, six miles to the north-east of Ypres, beginning at 19.30. The three single-seaters encountered 20 enemy fighters half an hour later. Second Lieutenant F Rose (B1551) reported that the German machines were some 2,000 feet higher and one or two seemed to dive in front of the Nieuports as bait. The Nieuports, led by a 24-year-old Australian, Captain Alfred Seymour Shepherd (B1504), circled, not falling into the trap, and tried to gain more height. However, the Albatros pilots came down and a fight started. Shortly afterwards Rose saw Shepherd below him, easily picked out by his leader's streamers, going down out of control over Poelcapelle with two German fighters on him. Rose dived to help but it was too late, for although he drove one Albatros away, he saw no more of the streamered Nieuport XXIII.

There were a number of British aircraft lost in this general area at this evening period: Alfred Niederhoff of Jasta 11, a 'Sopwith' at 21.10 (G) over Zonnebeke; Oskar von Boenigk of Jasta 4, a 'Sopwith' north-west of Tenbrielen at 21.00; while Kurt Wüsthoff, also of Jasta 4, claimed a 'Sopwith' at 21.10 south-west of Becelaere. Probably Niederhoff is the nearest to time and place, Jasta 4 claiming too far to the east and south-east of Ypres. Niederhoff achieved his sixth victory this evening, his seventh and last two days later. He fell in combat on the 28th, oddly enough to 29 Squadron's Nieuports. Shepherd had the DSO and MC and had become a flight commander just a week before his death, his DSO being announced just five days prior to his fall.

Black Sunday
Sunday 22 July 1917 was to see the loss of four British aces, two RFC, two RNAS, with another RFC ace wounded and a fourth forced to land. Without trying to decry the achievements of early British aces, it is nevertheless a fact that decorations were a little more liberally awarded in late 1916, early 1917 than was the case later. In the early days of air fighting, the bringing down of hostile aircraft was quite a feat, and it was new. Once it became commonplace, awards to victorious pilots were not so easy to come by. It was the same with pilots who brought down the first few German airships. They received Victoria Crosses, while later a DSO seemed sufficient.

Captain William Arthur Bond was among the early air fighters (like Shepherd noted above) to gain a DSO and an MC with Bar for five victories in less than a month. Billy Bond, a newspaper corrrespondent in peacetime, was an old married man of 27 in July 1917, and had served in the King's Own Yorkshire Light Infantry before transferring to the RFC. His wife later wrote a book about his flying life, *An Airman's Wife* (Herbert Jenkins Ltd, London, 1918).

Flying Nieuports with 40 Squadron his claims had been over a two-seater and four Albatros DIIIs between 10 May and 9 June. He had just been home on leave and returned to be made a flight commander.

On this Sunday he and three other pilots had flown to the Squadron's advanced landing ground at Mazingarbe. Bond led off a patrol at 06.05 with Bill MacLanachan, Bert Redler, Bill Harrison, John Tudhope and H A Kennedy. Over the Front Bond saw a balloon near Vitry and dived to the attack but broke off as enemy aircraft were spotted over Sallaumines. As they climbed to regain their lost height, anti-aircraft shells began to explode amongst them. MacLanachan's machine was rocked by a near miss and when he righted himself, he saw that Bond's machine had disintegrated and pieces were falling earthwards.

There does not appear to be any German AA units claiming Nieuport Scouts shot down, which is strange, and it has been suggested that a claim for a Nieuport by a two-seater crew of FA235, Unteroffizier Beyer and Leutnant Ebert, was that flown by Bond. Their victory went down at Loison, just east of Lens, which isn't far from Sallaumines. However, MacLanachan, in his book *Fighter Pilot* (Routledge and Son, 1936), notes that Bond's machine was blown apart by AA fire, and a report from a British AA battery says a Nieuport was seen falling over Sallaumines after its right wing was blown away by AA fire. Lieutenant Kennedy, however, recorded seeing Bond's machine spin down out of control after engaging three enemy aircraft, but does not mention anti-aircraft.

Later this July morning, four Triplanes of 10 Naval Squadron took off around 07.30, consisting of Flight Commander J E Sharman (N6307), Flight Lieutenant J A Page (N5478), Flight Sub Lieutenants C H Weir (N5379) and G L Trapp (N5354), to patrol around Ypres and Messines. Between 08.00 and 08.30 they encountered a large patrol of enemy aircraft and during the fight, in which some DH5s became involved, two German aircraft were seen going down out of control, one of which crashed. Strangely enough, Sharman was seen going after one German, then, just like Bond, he appears to have been hit by an AA shell and his machine broke up. Page apparently was the pilot who downed the two Germans, as Sharman was not seen to do so and neither Weir nor Trapp made claims, but Page himself failed to return.

Once again there is no apparent AA unit making a claim for a Sopwith Triplane, although one unit did claim a de Havilland east of Hooge. This appears to be a 32 Squadron machine (A9412). Otto Brauneck of Jasta 11 claimed a Sopwith east of Kortewilde, but this was timed at 11.25 (G), two hours later than Page was lost (08.15/09.15 (G)). There are no other more obvious claimants, so perhaps the German record of time is an error. Nor does there appear any obvious German losses to substantiate the claims by others that Page had downed two aircraft, therefore no 'missing' German might have been responsible for downing Page. Brauneck himself was lost on 26 July, thus this 22 July victory was his tenth and last.

The fact remains, however, that 10 Naval had lost two successful fighter pilots, Sharman (DSC and Bar and CdeG) and Page, both aged 24 and both Canadians, with eight and five (perhaps seven) victories respectively. John Sharman was flying 'Black Death'.

Captain Geoffrey Hornblower Cock MC flew Sopwith 1½ Strutters with 45 Squadron in 1917. In fact he and his observers claimed 13 official victories (unofficially more) between April and 22 July to be the highest scoring 'Strutter' pilot in the RFC. Cock was 21 and had been with the Artists' Rifles OTC prior to joining the RFC and had been with 45 in France since autumn 1916. On this Sunday he was part of a photo-recce mission of eight Strutters, tasked with taking photographs between Menin and the front lines. They had left their base at 10.20 and no sooner had they begun taking their pictures than they were engaged by 25 to 30 Albatros Scouts.

According to those who returned, the Germans appeared to fire an explosive shell high above the Sopwith formation, its smoke expanding to an enormous size which immediately attracted the German single-seaters, which dived into the middle of the two-seaters and completely overwhelmed them. Two of the Sopwiths were hit, one (B2576) flown by

Cock and his observer of the day, Second Lieutenant Maurice C Moore, who claimed one of the attackers in flames. The other, A1032, went down in flames, credited to Leutnant Niederhoff at 11.30 (G), south-east of Zonnebeke. Also in the fight, Lieutenant J C B Firth (a future Camel ace) had his observer, Second Lieutenant J H Hartley, killed, while the Sopwith's fuel tank was shot through. Chased by four Albatros Scouts and using petrol from his gravity tank, John Firth just managed to scrape back over the front lines.

When he returned from PoW camp on Christmas Day 1918, Cock stated: 'Whilst on a photo patrol about thirty enemy aircraft attacked; my engine and petrol tanks were pierced and we were followed down by five enemy machines. I came down at Warneton (west of Comines).' Cock retired from the RAF as a Group Captain in 1943. Cock was claimed by Oberleutnant Willi Reinhardt, the German's first confirmed victory, although he would go on to score 20 in all, win the Pour le Mérite and lead Jagdgeschwader Nr.I after Manfred von Richthofen was killed. He died in an aircraft crash on 3 July 1918.

In 1967 Geoffrey Cock told me:

> 'I was soaked in petrol from the riddled fuel tank and never knew why the machine did not catch fire. The left side of the rudder pedal was shot away by an explosive bullet, a splinter going into my ankle. It was later taken out with a penknife.
>
> The 1½ Strutter was a good aircraft but the front gun was quite useless. I was, however, lucky in getting the Ross interrupter gear on trial and this speeded up the rate of fire of the front gun to almost normal rate and I got most of my victories with it. I carried out 97 jobs over the lines and with my observers, shot down 19 German aircraft, of which 15 were certainties.'

The other two notable casualties this Sunday were Lieutenant David Langlands of 23 Squadron and Captain W E Young of 19 Squadron, both Spad pilots. Langlands, from Orpington, Kent, was 23 and had scored five victories during July. During a late afternoon OP, 23 Squadron were in a fight with Jasta 12, and Langlands' Spad (A6709) was badly shot about by Leutnant Bernhard Knake. Langlands came down just inside British lines, west of Hulloch, his radiator and tank shot through. This was the German's second kill. He would gain one more, a balloon claimed on 23 July, but would die in combat over Epinoy on the 24th.

Wilfred Ernest Young, who gained his third victory this day, was also wounded in the leg during an OP over Roulers–Menin (in Spad B1628), force-landing at Abeele aerodrome. He would go on to claim 11 victories, win the DFC, and command No.1 Squadron in 1918.

CHAPTER FIVE

July – September 1917

Otto Brauneck falls

As we read in the last chapter, Jasta 11's Otto Brauneck had brought his score to ten on 22 July. This 21-year-old from Sulzbach had seen service in Macedonia with a two-seater Feld Flieger Abteilung 69, then as a single-seat fighter pilot with Jasta 25. With seven confirmed and at least five more unconfirmed victories, he was posted to the Western Front in April 1917 to join von Richthofen's famous Jasta.

Just four days after downing his tenth kill, he intercepted Sopwith Camels of No.70 Squadron near Polygon Wood. Leading the Camels was a 20-year-old Londoner, who was actually born in Margate, Kent, Captain Noel William Ward Webb MC. Noel Webb had flown two-seaters too, FE2bs, with 25 Squadron in 1916, becoming an ace before going on to single-seaters and joining 70 Squadron towards the end of June 1917. He had now brought his score to eight. He had in fact been the first RFC Camel pilot to down a German aeroplane, on 12 July, which landed in British lines where its crew was captured.*

On the evening of 26 July, Webb led his Flight out at 14,000 feet, arriving over the lines at 18.45, seeing about 20 hostile machines over the German side. Not wanting to cross over at a disadvantage in numbers, he searched about and before long several other British scouts turned up, SE5s, Pups and Spads. With this back-up, Webb crossed above the trenches to do battle. The SE5 machines were in fact those of 56 Squadron. They had taken off at 19.00 hrs, six machines in all, Captain Prothero (A8925), Gerald Maxwell, Ian Henderson (son of General Sir David Henderson), Len Barlow, Robert Sloley and Eric Turnbull. These, together with 70's Camels, Spads of 19 Squadron and DH5s from 32 Squadron, got stuck-in and created perhaps the biggest dog fight of the war thus far. There were also some Naval Triplanes about. In his subsequent report, Noel Webb wrote: 'Early in the operation I saw a red-nosed SE5 diving on an enemy aircraft. The pilot seemed to me to dive

---

* The first actual Camel victory was scored by Flight Commander A M Shook of 4 Naval Squadron RNAS, based at Dunkirk, on 4 June 1917, one EA out of control; next day Shook crashed a German scout on the beach between Nieuport and Ostend.

his machine over the vertical and then both planes on one side folded back and the machine descended in a spinning nose-dive.'

This was Philip Prothero, a Scot from Ayrshire, who always flew in his kilt, who was 23 and had scored eight victories. Several pilots saw a machine break up, some saying it was an SE5, others an Albatros. Webb certainly attacked an Albatros in company with an SE5 and a Pup, and believed it had been badly hit. Webb continued:

> 'After this I noticed an SE5 rather far over by himself. I therefore went over to him for support. There were about six enemy aircraft just below me and on the way back to the lines I dived on the leading machine, letting off a burst of about 50 rounds. I saw the EA wobble and then fall plane over plane and finally spin. Later I thought I saw this machine crash on the ground. I then collected my formation and returned home.'

The SE5 pilots claimed one Albatros out of control, with Maxwell claiming a two-seater driven down. No.19 Squadron had lost one of their Spads. Webb (B3756) claimed his victim went down east of Zonnebeke at around 19.30. Brauneck crashed south of Zonnebeke at 20.45 (G).

However, Jasta 27 appears to have been in this fight too, and Vizefeldwebel Alfred Muth claimed his first victory, reported to be both a Sopwith and an SE5, south of Becelaere, timed at 20.40 (G). Leutnant Hans Helmigk was also shot down and crash-landed without injury. Whether Muth originally misidentified the 19 Squadron Spad for a Sopwith (the Nachrichtenblatt says Sopwith) and later historians changed the Sopwith for an SE5 to comply with Prothero's loss, is unclear. It fell south of Becelaere, so the area is right. From Webb's combat report it certainly appears that Prothero's machine simply lost its (starboard) wings during the dive into the attack. Some of 56 Squadron also reported a red-nosed machine breaking up, one not being certain of its type or nationality as it was already in pieces. Was this Prothero's machine, which seems likely, or Brauneck who, being a Jasta 11 pilot, would have had some red on his machine?

This is not as clear-cut as some would have us believe. My view is that Prothero's machine broke up in the dive, and that Muth misidentified a Spad for a Sopwith, which later historians changed to an SE5 to line up with Prothero's loss. Webb was certainly involved in two German fighters being shot down, but whether Brauneck was the one he attacked in company with an SE5 (Barlow?) and a Pup, or the 'leader' he attacked afterwards which he thought he saw crash, is unclear. Either one might have also been Jasta 27's Leutnant Helmigk. As Brauneck may have been leading a group of Jasta 11 pilots, the second claim seems the more probable.

Webb himself only lasted until 16 August, and Muth was killed on 5 September having scored a second victory by that time. Barlow's out of control victory seems to have been upgraded to a destroyed, bringing his score to six. He would raise this to 20 by the autumn only to die in a flying accident on 5 February 1918 in England.

Kroll's lucky escape

The next day (27th) another German ace went down, but this time he survived. Heinrich Kroll had just joined Jasta 24s after gaining five victories with Jasta 9. He had shot down a 56 Squadron SE5 on 20 July (Robert Jardine, who had made the head-on attack on Göring on 16 July) to mark his début with his new unit.

Kroll met Spads of 19 Squadron this evening and was hit by fire from one flown by Captain Clive Wilson Warman, an American from Norfolk, Virginia who had joined the RFC via the Canadian infantry. This was to be his third of an eventual 12 victories. Kroll recorded:

> 'I was shot down in flames but not injured – machine destroyed. I attacked ten Spads and brought one out of the formation and circled with him from 4,000 to 2,500 metres. Then he suddenly turned tail and went on the defensive. He shot my inlet pipes and induction valve, resulting in a fire in the carburettor. I immediately turned off the petrol, switched off the ignition and dived steeply. The machine smoked and burned, my face full of fuel and oil, and I could not see through my goggles; tore them off at 800 metres and looked for a place to land. With the prop stationary I pulled over a row of trees, under a high-tension cable, took some telephone wires with me, and landed in an open field where the machine came to rest on its nose; it was completely broken.'

Kroll's escape from an up-and-coming ace was fortuitous. He crashed at Beythem in Albatros DV 2075/17, north of Menin. He went on to claim a total of 33 Allied aircraft until wounded in combat on 14 August 1918, having in the meantime been awarded the Pour le Mérite. Clive Warman was to win the DSO and MC but was wounded in August 1917 and saw no further action. He died in an air crash in June 1919.

The month of July ended with three aces shot down on the 28th. As previously mentioned, Alfred Niederhoff of Jasta 11 was shot down by 29 Squadron. The Nieuports were patrolling west of Terhand at 11.00 hrs and ran into five red-painted Albatros Scouts. The British pilots claimed two destroyed and one out of control, Second Lieutenant C W Cudemore seeing his victim break up in the air before crashing. Niederhoff fell near Becelaere at noon (G), Charles Cudemore claiming his kill over nearby Gheluvelt at exactly 1100. He was flying Nieuport B1609 and this was his third victory of an eventual 15, flying with 40, 29 and later 64 Squadrons.

The Black Flight of 10 Naval Squadron ended its unhappy month with the loss of Ellis Vare Reid (N5483), another Canadian, who had gained 19 victories since 1 June 1917. His DSC was announced after his death. The Naval pilots had been flying an evening OP and got into a fight with a dozen German fighters. One was seen to fall in pieces which, as no one else made a claim, may have been shot down by Reid. Reid himself was last seen at about 20.00 following the engagement, in a clear sky, between Dadizeele and Roulers at 12,500 feet. It is believed he was shot down by AA fire. KFlak 41 reported shooting down a Triplane near Armentières, which seems a little too far south, unless Reid strayed this way looking for enemy aircraft. But it was the only Triplane lost this day, and the only one claimed by the Germans other than a Jasta 12 claim over 8 Naval earlier that morning.

The third lost ace was Laurence Minot of 57 Squadron, mentioned previously following the downing of Dossenbach on 3 July.

Two aces meet
The other Sopwith Triplane shot down on 28 July had been flown by Flight Sub Lieutenant E D Crundell, 8 Naval Squadron, who had just two victories, but who would end the war with seven. He survived the shoot down, by the German ace Adolf von Tutschek at 07.30 at Méricourt, on the British side of the lines – the German's 19th victory. That same 28th, Tutschek shot down another future ace, Lieutenant J H Tudhope of 40 Squadron, also into the British lines, near Lens. Tudhope had yet to begin his scoring, but would achieve ten by April 1918, at which time he was rested from operations. Eric Crundell once told me:

> 'Booker and I were fighting about ten Albatros scouts and two-seaters, Booker chasing one with me going for one behind him, but then Germans were behind me. In all the excitement I accidentally knocked the engine blip switch and my engine stopped, just as I was in a climbing turn. My Triplane flicked into a spin and went down which caused a problem as I had never been in a spinning nose-dive before and was wondering if I could recover.
>
> With the Germans still with me I decided to let my Triplane spin down in the hope they would think I was finished and leave me alone

and in consequence I was quite near the ground when I found I was indeed able to regain control. I got back over the trenches at very low level and flew back to base. My machine was riddled with bullet holes.'

Tutschek was to make a name for himself in the German Air Service, achieving a total of 27 victories and the Pour le Mérite before his own death in combat in March 1918. He also commanded Jagdgeschwader Nr.II from February 1918. He had commenced his scoring with Jasta 2 in 1917 but most of his kills were with Jasta 12.

On 11 August 1917, von Tutschek, a 26-year-old Bavarian, met his equal in air combat, an exceptional Naval pilot by the name of Charles Dawson Booker DSC, mentioned above in the Crundell action. From Kent, 20-year-old Booker had lived in Australia until 1911, and had joined the RNAS in 1915. He began scoring with 8 Naval in January 1917 and by this date had 20 victories, virtually the same as Tutschek, who now had 23 victories, having downed two Bristol Fighters from 22 Squadron already on the 11th. Flying Triplane N5482, Booker had taken off from St Eloi at 18.15 with two other pilots to fly an OP to Acheville. They met and engaged Jasta 12 over that location. It was later recorded:

'21.00 hours: Twelve SE5s and Triplanes reported. Engaged SEs and one went down smoking, followed by Tutschek but he then saw a Triplane above him circling, seemingly waiting for the right moment to attack. Tutschek's guns then jammed and there was a hit in front of him and the radiator is ruptured. Hot water splashes in his face and he can't see a thing; his Albatros trails a white steam cloud.

He lifts his goggles and sees the Triplane coming for him, the pilot probably thinking he's on fire. Tracer bullets whistle about him and hits [are scored] all over his machine. Then he is hit a terrific blow in the right shoulder. This stick and his useless arm flop into one corner and the machine goes into a spin, and he passes out.

He comes round at 1,600 m, still spinning, followed closely by the Triplane, still firing. Taking the stick in his left hand he cuts the ignition; he is spitting blood too. He is also six kilometres over the British lines.

Another aircraft attacks the Triplane and it goes down. Von Tutschek get his engine going and heads home at 150 metres in the direction of Henin-Liétard, followed by British ground fire. He tries to land at Douai but passes out at ten metres but comes to on the ground: the aircraft has landed itself in a field.'

Tutschek was lucky to survive. He had first been hit by Flight Sub Lieutenant W L Jordan, a future 37-victory ace, who would share this kill as his fourth victory. Disabled by this attack, Booker had then pounced. A second bullet had shattered the lower part of Tutschek's shoulder blade without damaging his spine or lung. He was soon in hospital and sent the following telegram home: 'After shooting down two Englishmen, slightly wounded in the shoulder. Hope to be back on the job in four weeks.' In fact he was not to return to the Front until the new year.

Meanwhile, Booker, seeing his adversary going down, was then attacked by Jasta 12's Leutnant Viktor Schobinger, and undoubtedly saved his Staffelführer's life. Booker's Triplane was badly hit and he had to make a forced landing near Farbus, where it was completely wrecked. However, Booker climbed free without injury. It was Schobinger's second victory of an eventual eight. Two days later, Tutschek gave his saviour command of Jasta 12, a job he carried until knocked out of the war by a wound on 15 November.

Mannock and von Bertrab

In August 1917, the name of Mick Mannock was only known amongst his pals on 40 Squadron, but today he is known as perhaps the greatest patrol leader of the war and one of the RAF's top scorers. Older than most of his contemporaries, he was 30 by this time, he almost didn't get to the war, being interned in Turkey when war began, while working for a British telephone company. However, he was finally released and after a period with the RAMC, joined the Flying Corps, one of his instructors being James McCudden. He joined 40 Squadron as a replacement for casualties inflicted during Bloody April and by the beginning of August had scored five victories.

On the 12th, he engaged an Albatros Scout, flying Nieuport B3554, at 15.15 and watched it come down in a field south-east of Petit-Vimy. He later wrote in his diary:

> 'Had a splendid fight with a single-seater Albatros Scout last week on our side of the lines and got him down. This proved to be Lieutenant von Bartrap (sic), Iron Cross, and had been flying for eighteen months. He came over for one of our balloons, near Neuville–St Vaast, and I cut him off going back. He didn't get the balloon either. The scrap took place at two thousand feet up, well within view of the whole front. And the cheers! It took me five minutes to get him to go down, and I had to shoot him before he would land. I was very pleased that I did not kill him. Right arm broken by a bullet, left arm and left leg deep flesh wounds. His machine, a beauty, just issued

(1 June 1917) with a 220 hp Mercedes engine, all black with crosses picked out in white lines, turned over on landing and was damaged. Two machine-guns with one thousand rounds of ammunition against my single Lewis and three hundred rounds! I went up to the trenches to salve the 'bus' later, and had a great ovation from everyone. Even Generals congratulated me. He didn't hit me once.'

The German pilot was Leutnant Joachim von Bertrab of Jasta 30. He had scored four victories on 6 April in two separate engagements, a record at that time, and had downed his fifth on 15 May. He had gone for the balloon at Souchez at 16.00 (G) but had been intercepted by Mannock shortly afterwards. His distinctive DIII was purplish-black and apart from the white crosses, it also carried a comet insignia on the fuselage sides. It became G.60, the location being recorded as Farbus Wood.

The 14th of August saw Captain T A Oliver, 29 Squadron, shot down and killed in Nieuport XXIII B1557. Tom Oliver, from Loughborough, was 21, and went down under the guns of Oberleutnant Ernst Weigand of Jasta 10, crashing at Neuvekapelle, near Dixmude, at 10.45 (G). Oliver had scored five victories with 1 and 29 Squadrons. It was Weigand's second kill and he would gain one more before falling on 25 September over Houthulst Forest.

Webb and Voss
Noel Webb, who had shot down Brauneck the previous month, also met up with Jasta 10 on 16 August. The star of Jasta 10, part of von Richthofen's JGI formation, was the mercurial Werner Voss, at this stage the nearest rival to the Baron himself. Baron von Richthofen scored his 56th victory on this day, Voss his 37th.

Voss was 20 years old, and came from Krefeld. A former Hussar, he had joined the Air Service in 1915 after seeing action on the Eastern Front. He had joined Jasta 2 in November 1916, Jasta 5 in May 1917 and had only recently taken command of Jasta 10. Despite being on leave for most of April when a number of German fighter pilots ran up big scores over the RFC, Voss had built a steady score over the spring and summer of 1917, and had been rewarded with the Pour le Mérite.

Webb, flying Camel B3756, now with 14 victories and a Bar to his MC, led an OP between Poelcapelle and Becelaere, taking off at 18.05. He was last seen diving upon two German aircraft at 4,000 feet over Polygon Wood about 19.45. Voss claimed a Camel down at about 21.00 (G) near St Julien, on the front line.

Two-seater pilots also became aces, whether sharing the claims of their gunner/observers or not. Canadian Harold W Joslyn, an FE2 pilot with 20 Squadron, was killed on the 17th, brought down with his Scottish observer, Second Lieutenant Alexander Urquhart, in B1891. They were part of a Distant OP on this morning and ran into fighters near Halluin at around 10.15 am. Jasta 6 was one of the units involved, Leutnant Deilmann downing their FE over Zonnebeke at 11.10 (G), his sixth and last victory. Despite his six victories, he now left the Jasta, flew two-seaters for a while, then became an instructor.

On the same day an Australian fighter pilot with 8 Naval Squadron, Philip Andrew Johnston (B3757), was killed. Ostensibly he died following a collision with fellow pilot Flight Sub Lieutenant R S Bennetts (B3877) during a dog fight with Jasta 30. However, as Oberleutnant Hans Bethge of the latter unit claimed both Camels, perhaps he felt he was the cause of their collision. In any event he was credited with both, his 12th and 13th victories, timed at 19.05 south-west of Wingles. Bethge was the Staffelführer of Jasta 30 and would secure 20 victories before his own death in action on 17 March 1918 (see later). Johnston had scored six victories flying both Triplanes and Camels.

Another Australian, this time a two-seat FE2 pilot, Cecil Roy Richards MC, from Garvoc, Victoria, was brought down the next day, 19 August. Roy Richards had achieved 12 victories with his various gunner/observers, and had been in the fight in which Manfred von Richthofen had been wounded back in early July. After serving with the army at Gallipoli and France, Richards had joined the RFC and then operated with 20 Squadron during the summer of 1917. On this day, however, his FE was crippled by Leutnant Ernst Hess, of Jasta 28w, and he and his observer, Second Lieutenant S F Thompson, were taken prisoner (B1890). They were flying a photo-recce sortie for XI Wing that afternoon, Hess claiming them down at Quesnoy, east of Wervicq at 18.00, the German's ninth victory. Hess would go on to score 17 victories before a Frenchman killed him on 23 December (see later). On his return from captivity, Richards recorded:

> 'Whilst on a photo-recce, a burst of "archie" hit the prop. I turned for the lines but owing to strong winds, it was impossible to reach them. Attacked an EA passing in front of me and was attacked by three more, who shot away all my controls and I was forced to land.'

Another successful two-seat pilot, this time flying BF2b machines, Canadian Captain Carleton Main Clement MC CdG, aged 21 (who has been mentioned earlier), was lost on the 19th. Flying A7172 he and his observer, Lieutenant R B Carter, were flying an OP over the 5th Army Front this evening. They were seen to be engaged at 19.35 but failed to return. They fell at Langemarck, claimed by AA fire from Flakzug 99, commanded by Leutnant Böhner (although this unit claimed Sopwiths – two of them). With 14 victories he was the most successful pilot on the Squadron at this time, and a telegram was received by the Squadron from no less a person than General Hugh Trenchard, deeply regretting Clement's loss.

A two-seater bites back
Oberleutnant Eduard von Dostler, commander of Jasta 6 and holder of the Pour le Mérite, brought down his 26th opponent on 18 August. Three days later he met his match, but not in a fashion he would have dreamed of.

On the morning of the 21st, an RE8 crew (A4381) of 7 Squadron RFC, Lieutenant Norman Sharples and Second Lieutenant M A O'Callaghan, took off to fly a photo Op over the Frezenberg–St Julian area of the front lines. Just after 11.00, the RE8 crew saw six light-coloured Albatros Scouts with silver prop bosses. Four of these came down on them, two on their tail, with one either side. O'Callaghan opened fire from 100 yards and one of the two rearward machines, the one on the starboard side, burst into flames and immediately nose-dived. He had only fired 55 rounds before his gun jammed, but he had been on target. Some bullets from one of the German fighters had hit an aileron wire, but Sharples was quickly diving for home and did not see if the burning Albatros crashed. But it had, and confirmation soon came from the Headquarters of the 143rd Infantry Brigade, who reported the Albatros had crashed at map reference C11.b.63 at 11.10, right up by the front lines.

Two days later the Germans dropped a message, requesting information on their lost ace, who they said had fallen about noon on the 21st. The British in turn sent word to the Germans, picked up and relayed to the relevant authorities. This resulted in a report:

> 'After a telephone communication by the Commander of Aviation, 4th Army, the English RFC has sent word concerning Oberleutnant Dostler, that particulars of his fate cannot be produced. It is only known that on 21 August 1917, at 11 am English time [12.00 German time], a German aeroplane was brought down by an English pilot in the vicinity of Frezenberg, and it probably lies in the forward-most German lines.
>
> Time and location in accordance with the above-mentioned particulars correspond with Oberleutnant Dostler's air combat at that

time. During the following night and the next morning, heavy British artillery fire lay over the site of the downed aircraft.'

Dostler, a Bavarian, was 25 years old and had had a long career in the army and air service. Norman Sharples was killed in action on 20 September 1917, but O'Callaghan was still on the Squadron in October 1918. He survived the war, but only just. On 8 October, he and his pilot, Second Lieutenant J Graham, were attacked by Pfalz Scouts over Moorseele. He shot down one attacker but was then wounded in the foot. Their RE was badly shot-up and crash-landed in no-man's-land, Graham breaking his nose in the pile-up, O'Callaghan sustaining a head injury. The two men lay in a ditch till after dark then got back to the British trenches. They were taken to a command post, and were told they had seen the Pfalz dive into the ground.

Another German commander falls
At the beginning of September, the Germans lost another Staffelführer, this time the leader of Jasta 28w, Oberleutnant Otto Hartmann, a 28-year-old Württemberger, who had also a long military career behind him.

It is not entirely clear who shot him down. We know that he was in combat north-east of Dixmude at 08.15 on the morning of 3 September, coming down at Kortewilde and dying of his injuries on the 7th, four days later. The most likely candidates are Lieutenants R E Dodds and T C S Tuffield (A7222) of 48 Squadron, flying escort to a bombing operation.

The area is right. Dodds and Tuffield, escorting DH4s, were above Keyem, north of Dixmude and were attacked by two Albatros Scouts on their way back. As the first one attacked, Dodds manoeuvred and brought the front gun of his BF2b to bear, and as he passed beneath their right wing, Tuffield also fired and the German machine appeared to go down out of control. Meanwhile, the second Albatros got on to their tail and slightly above. The British crew turned and the two aircraft fired at each other head-on, whereupon the Albatros started down, leaving flame and smoke and this too appeared totally out of control. Kortewilde is west-north-west of Dixmude, Keyem just further north. The Naval DH4s from 5 Naval Squadron had been attempting to attack Varssenaere airfield, but they were mostly scattered by the attacks of German fighters. Nevertheless, Jasta 20 had a pilot killed in the bombing and two wounded, one fatally.

The Camels of 3 Naval Squadron were also in a scrap in the general area, Lloyd Breadner (B3782) and Bill Chisam (B3909) (an ace and future ace respectively) claiming Albatros Scouts destroyed and out of control at Bilhutte and Stelhilde around 07.30. Marine Feld-Jasta Nr.I also had one of its pilots shot up and wounded in the general area.

Jasta 7 was the unit in action with the Naval pilots (and some French Spads from Spa 48), Josef Jacobs seeing an Albatros 'explode'. This appears to have been Hartmann, but Breadner's confirmed kill is supposed to have lost wings from one side. Whatever the outcome, Otto Hartmann had been knocked out. He had achieved seven victories, five while leading Jasta 28.

The great Heurtaux is wounded
Capitaine Alfred Marie Joseph Heurtaux had scored 21 victories during 1916–17 and become one of the heroes of France. All his claims had been with the 'Storks Group', of which his Escadrille N3 was part. However, he had been severely wounded in combat on 5 May 1917 and had only returned to his unit on 7 August.

He had not added to his score by 3 September, the day which saw his last action. Leading his Escadrille on a patrol that evening, they were spotted by two pilots of Jasta 7, Kurt Schönfelder and Otto Kunst, near to the lines. Leutnant Kunst fired into the leading Spad and saw it fall away towards the French lines. Heurtaux was wounded but got down near Pervyse, but without official witnesses Kunst's claim went unconfirmed. Whether Heurtaux was surprised or just out of practice is open to question, but Kunst had only been with Jasta 7 for one month, and had just one victory to his name, scored back in June with Jasta 22. He did not score any others before going to the Junkers Aeroplane Company in March 1918, yet he had knocked out the great French ace.

Heurtaux recovered from his wound but did not see further combat. Instead he was sent to America where he lectured to the future generation of American fighter pilots.

French retaliation
The French were able to retaliate the next day, 4 September; Erich Hahn of Jasta 19 was attacked and shot down north-west of Beine in the Champagne region at 19.35. Just like Heurtaux the previous day, Hahn had not scored for a while, his sixth and last victory having been claimed back in April. The French ace Georges Madon of N38 claimed an Albatros over Nogent l'Abbess this day, his 15th victory. He would go on to score 41.

Clive Collett's last air fight
Clive Franklyn Collett was a New Zealander, from Blenheim, and had just passed his 31st birthday. He had been flying since 1915 and had risen to Captain while serving with 70 Squadron, one of the first RFC units to equip with Sopwith Camels. Between 9 July and 8 September he had

scored eight victories and been awarded the MC and Bar. On 9 September he would claim three more but receive a wound that put him out of the war.

It was a late afternoon patrol, Collett in B2341. In the hour between 17.00 and 18.00, he had three fights over the Gheluvelt–Houthulst area, claiming two two-seaters and an Albatros DV. The last fight, with the Albatros, occurred at 17.50. And if it is difficult to be certain of who shot down Otto Hartmann, it is equally difficult to conclude who wounded Collett.

This final fight was in the Houthulst Forest area, and at this general time, Jasta 27 and 35b had combats with Camels. Collett had re-formed with his unit after the earlier scraps with the two-seaters and was patrolling once more towards the Forest. He spotted two enemy machines beyond the wooded area, towards Roulers, but then heard machine-gun fire, turned and saw his formation beginning a combat with several German fighters. Collett got on to the tail of one and fired a burst into its fuselage at short range. He followed it down and saw that the German pilot was making an effort to land. Collett shut off his engine and flew straight at the Albatros, firing a long burst into him as he was now on the ground, the German machine bursting into flames. The pilot was undoubtedly Leutnant Karl Hammes who had been with Jasta 35b for little under a month. He had turned over on trying to land and his Albatros, 2336/17, was smashed and written off. Hammes himself suffered wounds to both arms and ankles, stomach and upper left thigh and right toe. He was taken to Feld Lazaret (hospital) 502 at Lichtervelde.

Hammes later reported shooting down a Camel at Stadenberg before he was hit. Other claims for Camels at this time were Jasta 4's Oskar von Boenigk, at Poelcapelle, 19.05 (G), and Fritz Kosmähl of Jasta 26, at Westroosebeke, 19.00 (G). Karl Hammes was the only reported casualty on the German side this date, although 70 Squadron claimed another Albatros DV crashed. The only Camels lost were those of Lieutenant N C Saward, PoW (B3928), and Second Lieutenant H Weightman, (B3916), who came down in the front lines, was wounded, but got back.

Confirmation was not immediate for the German pilots. Kosmähl's was confirmed almost at once, his seventh kill, but Hammes' claim was not upheld for some while, but it was later, his fourth and last. It may be that von Boenigk's 'Sopwith' had been a Triplane of 1 Naval, and if so then Kosmähl and Hammes got the 70 Squadron Camels. As it was, Collett was then attacked, hit and wounded by another Albatros but limped back. As he did not come down, the German pilot was unable to put in a claim, but Ludwig Hanstein of Jasta 35b did attack a Camel and may well have inflicted the injury to the New Zealander. Hanstein had ten victories as of this date and would score a total of 16 before his own death in March 1918. Collett returned to England and upon recovery became a test pilot but on 23 December 1917, while flying a captured Albatros Scout near Edinburgh, he had a problem in the air and crashed to his death over the Firth of Forth. As for Karl Hammes, he became a star of the Viennese opera in the 1930s, but was killed in action over Poland in 1939.

The French lost another ace on 10 September, this time Jean Georges Fernand Matton, of Escadrille N48. This 28-year-old had won the Légion d'Honneur with the Dragoons before joining the air service. Flying Nieuports, first with N57 and then N48 he had gained nine victories before being killed. He was shot down in flames at Bergues in the Flanders Sector, while on a four-man patrol that engaged Albatros Scouts south of Nieuport. His victor was Josef Jacobs of Jasta 7, his victim falling near Keyem at 19.05 (G).

CHAPTER SIX

September – November 1917

The ace who flew too high

September 1917 was proving a savage month for the flying aces. On the 11th four would die, the most famous being the great French pilot Georges Guynemer. Georges Marie Ludovic Jules Guynemer, the 22-year-old Parisian, was the darling of France. Despite poor health, he had accounted for 53 German aircraft since the summer of 1915 and won every decoration his grateful country could bestow upon him. In addition he had claimed 35 probables. He had flown Moranes and Nieuports but his famed Spad with the red 'stork' marking of Spa3, part of the Cigognes Groupe, and his personal 'Vieux Charles' below the cockpit, was well known, not only at the battle front but in the newspapers and journals on the home front.

Like a number of top aces, his death was shrouded in mystery. On the morning of 11 September, Guynemer had taken off at 08.25 to fly a patrol over the Front in company with Sous-Lieutenant Jean Bozon-Verduraz, who had yet to score his first victory but who would end the war with 11. Crossing the lines near Bixschoote, heading for Langemarck, Guynemer found a two-seater and attacked but it evaded not only him but the attentions of both Frenchmen. The German began to spin away in an attempt to escape, Guynemer following, trying to ensure he wouldn't. Bozon-Verduraz watched but eight German scouts were now above and in keeping his eye on them, Bozon-Verduraz lost sight of his leader. The German fighters did not intervene and soon they flew away but of Guynemer there was no sign. Bozon-Verduraz eventually had to return home, expecting to meet Guynemer there, but he did not return.

It was some time before it was announced that Guynemer was missing and the Germans did not, as one might have thought, announce the fall of the leading French ace. It seems he fell into the front lines amid an artillery barrage, and simply disappeared in the explosions. This being the case, it took some time for the fifth victory of Leutnant Kurt Wissemann of Jasta 3 to be confirmed, and there was no identification of the victim due to the circumstances of his fall. Thus when it was finally admitted that the Frenchman was missing, the Germans searched their records to reveal their Jasta 3 pilot had downed a Spad XIII on this date and it must be the missing ace. As it seemed to the French that the Germans took so long to report a victorious pilot, they believed it had all been made up, but it was simply that there had been no idea of who the downed pilot had been. There is a note that Guynemer's body was located by German troops but it was left and subsequently lost in the barrage, the fate of not a few airmen, not to mention thousands of foot soldiers. The pilot credited with the victory had been an observer with FA250 before pilot training. Guynemer, who Wissemann reported he downed over Poelcapelle at 10.30 (G), was to be his last. The German was himself shot down and killed before the month was out.

An interesting aside to this episode was discovered by this author in his early researches into the WWI aces, in the late 1960s. It seems that when the French reported Guynemer missing, the RFC must have given their own records a glance, and found that their own 20 Squadron had claimed an unidentified fighter shot down in the area. Reference to it was noted in the RFC Communiqués, information for which would have been gathered prior to the announcement of the loss, and printed up. When I tried to check the combat reports for 20 Squadron, they were missing from the file (and this was long before the famous, or infamous, thief began stealing irreplaceable documents from the Public Record Office). A check on the Squadron Record Book also found a blank space, and it seems the combat reports and the record book page had been spirited away – at the time. Did, I wonder, someone begin to panic, thinking that the RFC had downed the great Frenchman? If so, how they must have cursed the fact that the Communiqués had already been printed and distributed.

However, while I personally tingled over this, I was relieved to discover some time later, as no doubt RFC HQ found out later, that their man had claimed the unidentified scout at 14.30 in the afternoon and not around 09.30 in the morning. Doubtless the French had not bothered to state the time that Guynemer went missing and one can imagine a discreet telephone call by RFC HQ to the French, saying how sorry they were, and by the way, could they tell them the time of his sad loss. Another odd thing was that Guynemer had been trying to achieve his 54th confirmed victory, and Wissemann's claim was Jasta 3's 54th confirmed victory.

For Guynemer, suffering ill health, death in combat was probably the way he would have wanted to go. Certainly when asked what other decoration his country could bestow on a man who had won them all, he is supposed to have answered, a wooden cross. And for the school children of the time, his loss was reported by saying their great hero had flown so high he had been unable to get down again.

Voss gets his 47th

Werner Voss had continued his steady scoring rate. Victory No.46 fell to him this September morning, a Brisfit; No.47 came in the afternoon. The Camels of 45 Squadron took off on patrol at 13.50, led by Captain Norman MacMillan. At 10,000 feet east of Langemarck they ran into five or six of the new Fokker Dr1 Triplanes, all buff-coloured (or so they were reported) and very quick and manoeuvrable. MacMillan dived on to a Triplane and firing from close range saw it wing over and go down in a spin. He then attacked a second Triplane attacking one of the Camels, following it down from 10,000 to 4,000 feet. MacMillan put three bursts into it from 50 yards and it fell over sideways and went down out of control.

MacMillan was credited with two Dr1s out of control, his sixth and seventh victories. He would score 11 in all. They were timed at 15.25. MacMillan knew he'd been up against an expert pilot, and believed the German must have wounded or killed the Camel pilot before he could intervene, for the pilot, Oscar McMaking (B6236), made little attempt at evasive action after the first attack. The German pilot, whose goggled eyes looked back at MacMillan as the latter tried to nail him, made a rapid movement and was soon out of the line of fire and going down. Voss got back, of course, and Jasta 10 lost no pilots this day, but they claimed the six-victory ace, McMaking. He crashed near St Julien at 16.25 (G).

No.1 Squadron lost two pilots on this 11th day of September, both aces. One was Captain Louis Fleeming Jenkin MC, aged 22, the other Lieutenant William Stanley Mansell, two weeks short of his 21st birthday and who had scored his fifth victory two days earlier. Jenkin achieved his 22nd victory on the morning of this day, a grey Albatros DV with light green wings, which he sent down out of control over Gheluwe at 08.35, flying Nieuport B3635. Mansell was flying on another patrol later in the morning, five Nieuports taking off at 10.43. The Nieuports headed for the patrol line of Ledeghem–Menin–Quesnoy and found two two-seaters at 11.00, but they rapidly drew away east. Two more were seen but they too became elusive. Then at 11.20 the British scouts came under AA fire south of Houthem and Mansell's machine (B3648) was seen to receive a hit. The machine's right wing was shattered and he went down in a spin, shedding pieces of wreckage. He was buried at Pont-du-Hem.

Jenkin flew another patrol at 15.45, six aircraft operating over the northern patrol area between Roulers and Menin. At 16.20 six enemy fighters were engaged by the Nieuports which had been joined by some Naval Triplanes. The Germans dived east. Then at 17.15, seven German fighters and two two-seaters were spotted over Westroosebeke and were engaged. One of the two-seaters worked its way onto Jenkin's tail but was driven off by another Nieuport, and Jenkin was last seen at 17.20, low down but under control. An hour later he had failed to return. He was claimed by Oberleutnant Otto Schmidt, the Staffelführer of Jasta 29, at 19.30 (G) over Bixschoote, the German's eighth victory. He would gain 20 victories at least by the war's end. Unlike his Squadron comrade, Jenkin has no known grave.

Little Wolff is killed
Among the high scorers of Jasta 11 and JGI was Oberleutnant Kurt Wolff. Slight of build he had nevertheless scored 33 aerial victories with Jasta 11 and Jasta 29 since March 1917. Holder of the Pour le Mérite, he had returned to Jasta 11 as commander on 2 July but was wounded in the left hand in a fight with 10 Naval Squadron on the 11th. He was away from his unit for exactly two months, returning on 11 September just as a couple of the new Fokker Triplanes had arrived, and were being shown off to von Richthofen, Voss – and Wolff. Within a couple of days, Wolff had shown his liking for the new, nimble Triplane, and flew FI 102/17 on the 15th. As fate would have it, he ran into 10 Naval – again!

An eight-man patrol from this British squadron had taken off at 15.15, led by Flight Lieutenant D F Fitzgibbon (B6202). The patrol picked up five DH4 bombers at 16.10 and went round Bryke Wood and back to the lines with them. At 16.30 over Moorslede, the Camels (10 Naval had exchanged their Triplanes for Camels by this time) saw four German scouts but no sooner had the Naval pilots shown an interest than they dived away. However, they had lost a bit of height and were then in turn attacked by five Albatros Scouts and four Triplanes. A general engagement followed in which one Triplane was hit and sent down out of control by Flight Lieutenant Norman Miers MacGregor, a Londoner aged 21. MacGregor got in really close to the Triplane – 25 yards – his tracers entering the Fokker but he then had to turn to avoid a collision. The Fokker nose-dived. Although he did not know it, he had downed Wolff, who crashed near the village of Nachtigall, north of Wervicq and was killed. It was MacGregor's fifth victory. He would score two more by the end of the year and receive the DSC. Wolff's Triplane was the first to be lost in the war.

Six days later, on the 21st, 48 Squadron lost its most successful pilot to date, Lieutenant Ralph Luxmore Curtis, flying with observer Second Lieutenant Desmond Percival Fitzgerald Uniacke. Both were from Essex, Curtis from Rainham, Uniacke from Upminster. By this date Curtis had achieved 15 victories, Uniacke 14, but oddly, neither had been decorated.

They had taken off at 08.00 hrs to fly on a combined OP and escort to bombers going for Ostend and were last seen about ten miles east of Roulers. Later a BF was seen going down in a spin ten miles north-east of Roulers. Curtis did not survive the day, but Uniacke became a prisoner. On his return from Germany in 1918, he reported:

> 'Escorting bombers over Bruges, a formation of eight enemy aircraft came up and three attacked. One EA forced down out of control and crashed. I was then wounded in the head. My pilot could not keep up with the Flight owing to the engine cutting out. Curtis was killed and aircraft totally crashed.'

Their BF2b (A7224) was claimed by Hermann Göring (Albatros DV 4424/17), leader of Jasta 27, downed at Sleyhage, west of Roulers, at 09.05 (G). It was his 14th victory of an eventual 22. He had taken command of Jasta 27 in May and would lead it until given command of JGI in July 1918. His combat report read:

> 'Shortly after 9 am sighted an enemy Geschwader of 14 aircraft (bombers), who were returning from Torhout in the direction of the front. I attacked with my escorts at an altitude of 4,500 m. I positioned myself close under the tail of an opponent and fired at him. I followed him closely with one of my other planes. Near Sleyhage, west of Roulers, the plane crashed. Pilot and observer were injured. Aircraft completely destroyed.'

Died of wounds
Offizierstellvertreter Fritz Kosmähl came from Leipzig and had started pilot training in August 1914. As with most successful fighter pilots, he began life as a two-seater pilot and became one of the first German airmen to receive the Prussian Member's Cross with Swords of the Royal Hohenzollern House Order two years later. By that time he and his various gunners had scored three aerial successes. Two more victories with another two-seat unit and he was an ace. He then became a fighter pilot, flying with Jasta 26 which he joined in August 1917. On 20 September he downed his ninth opponent. Two days later, the 22nd, he was over the Front mid-morning and ran into SE5s of 60 Squadron.

The British squadron were on a Northern OP, south-east of Zonnebeke. At 10.15 the SEs met six Albatros Scouts between Gheluvelt and Zonnebeke, just beneath the clouds. Captain R L Chidlaw-Roberts, aged 21, a former Sandhurst man, observer, and now fighter pilot, attacked the leader firing Lewis gunfire into him at 150 yards' range, and it dived away steeply and was not seen again. Half an hour later the SEs again engaged enemy fighters, five Albatros Scouts. Bob Chidlaw-Roberts (A8932) and Captain H A Hamersley (B523) attacked a green and black scout at close range and it slowly turned on to its back and spun down, apparently out of control.

One of these machines was flown by Kosmähl and he had been hit in the stomach coming down south of Poelcapelle, landing without further mishap. He was taken to hospital but died of his wounds four days later. Harold Hamersley was credited with one out of control victory, his second claim. He would score 13 by early 1918 and receive the MC.

Werner Voss
Captain Hamersley featured again in the early part of an action on 23 September, often referred to as the greatest dog fight of the war. On this day, the great Werner Voss, leader of Jasta 10, was on his last day of duty prior to going on leave. Indeed, his father and brother had come to the Staffel's airfield to return with him to Germany.

Voss had flown that morning, shooting down a DH4 as his 48th victory. Later that afternoon, he decided to fly one more patrol. Was he hoping to achieve a score of 50 before his leave started? Back in the spring Manfred von Richthofen had tried the same thing and scored four kills before a leave period began, raising his score from 48 to 52 on 29 April. Perhaps Voss, still von Richthofen's nearest rival, would have liked to have done the same, or at least to get two to make it a round 50. This may go some way to understanding why Voss did what he did this afternoon.

Voss was on his own, although he was joined by an Albatros Scout during the coming action. The SE5s of 60 Squadron were out this afternoon, Captains H A Hamersley and Chidlaw-Roberts being among them, the remnant of a larger patrol. They were bringing up the rear and saw some German fighters, and then Hamersley thought he saw a Nieuport Scout being attacked by an Albatros. No sooner had he gone to its aid, than the Nieuport turned, Hamersley realising it was in fact one of the new Fokker Triplanes. The expert Voss outmanoeuvred Hamersley and severely shot up his SE5a, forcing the latter to go into a spin. Chidlaw-Roberts tried to intervene, but Voss quickly shot his machine about too and he had to dive away. Voss may well have wished to finish off his two damaged opponents, but at that moment another formation of SE5s arrived – 11 machines of 56 Squadron, shortly before 18.30.

There followed a classic duel, the 20-year-old Voss in his Triplane constantly outmanoeuvring the British scouts. A red-nosed Albatros also joined in but it was soon out of it, leaving Voss almost in control of the whirling air fight, putting bullets into most of the SE5s and seeming to be invulnerable. Two of the SEs had to retreat, the aircraft damaged. But he was now fighting the cream of 56 Squadron, James McCudden, Gerald Bowman, Richard Maybery, Arthur Rhys Davids and Reg Hoidge, all high-scoring aces in their own right.

Finally Voss's yellow-nosed Triplane was hit and his engine stopped. The fight had drifted over British territory and Voss now turned to get back across the lines before he came down. The damage was done by Rhys Davids (B525), who then saw the Triplane, its engine off, gliding west. He says 'west' but one wonders if he meant east – or was Voss wounded and just going down? Rhys Davids attacked again, fired once, then twice, seeing the Triplane make a slight right-hand turn, still going down, then he overshot it, at 1,000 feet, and did not see it again.

Voss and his machine came down at Plum Farm, just north of Frezenberg at 19.35 (G). The German ace was dead, and was buried where he fell, on the British side of the lines. Like so many others, his grave was soon lost in the shelling and battles that were fought over the area. (The Menin Road Ridge actions were taking place at this time.) Voss had put bullets into every SE5 and had severely damaged four. He had so nearly achieved his 50, and so nearly emulated his rival Richthofen with four kills prior to his leave. His father and brother searched the empty evening sky west of Marckebeke in vain.

He could have so easily broken off the fight, for although the SE5 was faster than the Triplane, its manoeuvrability and his personal skill would surely have protected him. He was probably enjoying himself and may have been thinking hopefully that at least a couple of the SEs would have been seen by German observers to have crashed into the front lines. Final realisation of his over-exuberance only came when either his engine or fuel line was hit or perhaps a bullet had found him.

Indeed, the 19-year-old British pilot later lamented, when it was known that Voss had been killed, if only he could have brought him down alive. If he had seen Voss with engine off truly going west, as he said, then there might have not been any reason to 'finish him off'. But in the heat of action . . .! And the Albatros pilot? There was at least one, perhaps two, but one at least is known to have been Carl Menckhoff of Jasta 3, an 11-victory ace who would raise his score to 39 before being taken prisoner in 1918. He failed to score in this fight, not surprisingly, and was hit, also by Rhys Davids' fire, which brought his score to a round 20. Menckhoff crash-landed north of Zonnebeke.

Double header

The Bristols of 20 Squadron knocked down two aces on the evening of 25 September, although both men survived the experience. They were flying a Northern OP over Becelaere at 18.30, four BF2bs, and were attacked by six Albatros Scouts. Lieutenants Henry G E Luchford MC and R F Hill (B1122) were an experienced two-seater crew, Luchford having 14 victories at this time, mostly on FE2d pushers.

They claimed one attacker shot down east of Geluveld (also spelt Gheluvelt), while Second Lieutenant N V Harrison and Lieutenant V R S White (B1126) engaged another which had dived on them. This machine went down trailing smoke and flame. Two other crews claimed out of control victories. Jasta 8 appear to have been the attackers on this occasion and they had Leutnant Walter Göttsch shot down and wounded, not seriously, while Leutnant Rudolf Wendelmuth was hit and forced to land near Passchendaele with a shot-up engine. These two had 17 and ten victories respectively and would score more before they were both killed. Norman Harrison was killed too, in a flying accident in 1918. Victor White (MC & Bar) would be observer to Luchford in late 1917.

The Eagle of Lens

One is never certain how some airmen receive titles such as that of Oberleutnant Hans Waldhausen, whom history seems to have noted as being the Eagle of Lens. He was 22 when the war started and had been an army cadet for three years by then. With a Bavarian artillery unit when the war began he was wounded in September 1914, he then transferred to the air service the following year. At first with two-seaters, then single seaters, he joined Jasta 37 at the end of July 1917. Due to his rank, he was given acting command of the Jasta on 26 September, but the very next day was his last in combat. It had been a busy month for Waldhausen. He had downed his first opponents on the 19th, one confirmed, one unconfirmed. Victory No.2 came on the 24th, followed by a balloon on the 25th. Thus in just eight days he had scored three victories around Fresnes, Cagnicourt and Bethune, to gain his title, which is a little amazing.

On the 27th, his second day of acting command, he downed a balloon south-west of Roulette at 17.05 pm (G), and a RE8 just five minutes later over Farbus Wood, then another balloon at Neuville–St Vaast at 18.15. However, all this activity, especially around the British balloon lines, had attracted the attention of British fighters. Captain C D Booker DSC, flying a Camel of 8 Naval (B6227), was returning from a patrol, in company with two other pilots, and observed Waldhausen's Albatros, east of Lens. Booker thought the German seemed to be about to attack the balloon lines and he was right. He saw the German dart across and flame a balloon of the 37th Section, 3rd Company, 1st KB Wing and he caught Waldhausen as he headed east again. Booker fired a good burst into him and then the German was hit by another attack, this time by a Nieuport.

Booker timed his attack at 19.00, and so did Second Lieutenant J H Tudhope of 40 Squadron (B3617). John Tudhope, a South African, had just one victory to his name, but would end the war with ten. Earlier he had been sitting in his aircraft at the Squadron's advanced landing ground when he observed an Albatros Scout being fired on by British AA guns whilst approaching the balloon line. He immediately took off and headed straight for the German, whom he saw shooting at the gasbag. The Albatros then turned and Tudhope was then between it and the front lines. Tudhope attacked twice, also noticing a second Nieuport (sic) making an attack. Badly hit, the Albatros went down and crashed alongside the light railway station at Souchez, and was immediately surrounded by British troops. Waldhausen was captured, his brief period as the Eagle of Lens at an end. However, in the exchange, Waldhausen had put some shots into Tudhope's machine and he was obliged to make a force-landing, ending up in a shell hole, his Nieuport turning over badly damaged. Victory No.7 for the German, making four in one action?!

Waldhausen's Albatros DV (2284/17) was given the British serial for captured aircraft, G.74. It had a varnished plywood fuselage, which gave it a yellowish appearance, and on the fuselage sides it carried the German national markings and also a Star and Crescent motif. After the war he became a judge. Booker had gained most of his victories thus far in Sopwith Triplanes, this being his first with a Camel. Tudhope had now scored his second and final Nieuport victory; he would now fly SE5a scouts.

56 Squadron again
The experienced SE5 pilots of No.56 Squadron claimed another ace on 28 September, this time Leutnant Kurt Wissemann of Jasta 3. This 24-year-old from Elberfeld had been an observer prior to flying fighters, and had been wounded at the end of April 1917. His claim to fame, however, had been when credited with the victory over Georges Guynemer on 11 September 1917.

A romantic version of Wissemann's fall was that he was downed by René Fonck, who would end the war as the French Ace of Aces with 75 confirmed victories, but this is not so. No.56 were patrolling the Thourout–Ruddervoorde–Courtrai–Menin areas, having taken off at 16.30. Two enemy fighters were seen and engaged at 17.20 east of Westroosebeke. Reg Hoidge attacked the lower of the two, and it went down in a steep spiral and broke up, Hoidge watching the pieces fall into a wood. Gerald Bowman also saw the two scouts fly up through a gap in some clouds. His attack too sent one Albatros down, almost vertically for about 5,000 feet. Then something broke off and the whole machine disintegrated. Two Jasta 3 pilots went down, Wisseman killed, and once again Carl Menckhoff, who'd been with Voss in his last fight, crash-landed. However, Menckhoff again survived. Whether both British pilots saw the same machine break up low down is unclear, but 56 had got their men.

Captain Alwyne Travers Loyd (yes, one 'L') was also killed this day. A minor ace with 32 Squadron, he had gained his six victories flying FEs with both 25 and 22 Squadrons, then in DH5 machines of 32. 'Button' Loyd went down east of Ypres (A9211) shortly after 13.00, while on an OP, thought to have been hit by AA fire, whereupon his machine broke up. He was 23 and is buried at Lijsennthoek, Belgium. Rudolf Berthold, leader of Jasta 18, was also supposed to have downed a DH5 at 12.30 (G), south of Zillebeke Lake. Originally records noted his victory as over a Martinsyde, while two other pilots of the Jasta both claimed BF2bs at the same time and general area. No.20 Squadron lost two Bristols at 12.30, south of Wervicq.

56 Squadron come second
No. 56 Squadron were not so lucky on the first day of October 1917. 'A' and 'B' Flights had flown out at 16.30 hrs. B Flight attacked three two-seaters over Becelaere but they got away. McCudden and Rhys Davids lost contact, climbed to 11,000 feet west of Moorslede and flew beneath seven German fighters to try and lure them towards the lines and below A Flight, but the German pilots, led by one in a black and white Pfalz DIII, dived on them instead. Gerald Maxwell's A Flight joined in as did some BF2bs and then some more German scouts arrived, so that more than 20 German aircraft were involved with the SE5s and Brisfits.

The first bunch of German fighters belonged to Jasta 26. McCudden saw an SE5 circling inside four Albatros Scouts and then one Albatros turned inside the SE and, at 25 yards' range, shot the SE's tail off, the rest of the machine going down in a spin. The luckless pilot was 20-year-old Lieutenant Robert Hugh Sloley (A8928), from Cape Town, South Africa, who had achieved his ninth victory on 29 September. He was the only son of Sir Hubert and Lady Charlotte Sloley. He was claimed by Leutnant Xaver Dannhuber of Jasta 26, the German's sixth victory of an eventual 11. The SE came down at Westroosebeke at 15.50 (G). Jasta 26, of course, flew machines with black and white fuselages.

Second Lieutenant Colin MacAndrew, from Ayr, Scotland, had scored five victories flying BF2bs with 11 Squadron in the summer of 1917 but was then posted to fly DH4 bombers with 57 Squadron. It was not usual for such a change of type and role, and sadly MacAndrew did not last long with his new unit.

On 2 October, the Squadron mounted a raid on Abeele aerodrome, six aircraft led by Captain D S Hall/Lt E P Hartigan (A7568), soon after 13.30 that afternoon. Over Roulers the bombers met a large force of German fighters flying at 17,000 feet. The DH4 crews claimed two in flames and two out of control but three of the 'Fours' did not get home. One of these was crewed by MacAndrew and his observer, Second Lieutenant L P Sidney (A7581). Of the six men, just one survived as a prisoner.

They had run into Jasta 18, Rudolf Berthold, Leutnant Richard Runge and Leutnant Walther Kleffel claiming their 28th, eighth and first victories respectively. However, Kleffel was one of those claimed shot down, being severely wounded. It seems that Berthold brought down the 20-year old MacAndrew. He and his observer are buried at Harlebeke, Belgium.

56 Squadron lose another
Charles Hugh Jeffs became an ace on 29 September 1917, his fifth claim in less than a month. Jeffs was a Grimsby lad and had joined 56 in mid-August. He was part of a dawn patrol on this 5 October morning, flying B524, along the line Thourout–Ruddersvoorde–Courtrai–Menin. The Flight was engaged by German fighters over the Roulers–Menin road and Jeffs was not seen after the combat. Once again it was Jasta 26 that inflicted the loss, Oberleutnant Bruno Loerzer, the Staffelführer, sending Jeffs down at 08.20 (G) to begin 15 months as a prisoner of war. It was Loerzer's 13th victory; he would gain 44 by the war's end and lead Jagdgeschwader III.

Two days later, 7 October, 22 Squadron lost Lieutenant J C Bush MC, an ace with seven victories. A clergyman's son from Wiltshire he had flown FE pushers with the Squadron but did not achieve any victories until the unit re-equipped with Bristol Fighters. Jim Bush and his observer, Lieutenant W W 'Bill' Chapman were shot down and killed on another dawn patrol, flying an escort sortie in A7280 over the British 5th Army Front. They were last seen with four other machines over Kruiseik at 07.30. In the event, they were engaged and brought down by Leutnant Hans Gottfried von Haebler of Jasta 36, the first victory for this 24-year-old who had joined this Jasta at the end of the previous month. He had, however, a vast amount of experience of war flying, having been with a two-seater unit, following war service in the infantry. His first victory fell in the Menin area at 08.40 (G). In March 1918 he was brought down inside British lines and died of his wounds.

Dannhuber gets one more
Xaver Dannhuber of Jasta 26 claimed a Spad on 9 October 1917. However, it was not an uncommon event for pilots of both sides to misidentify the aeroplanes they were in combat with, a fact that has misled air historians for many years. It has been thought that his victim was a Spad pilot of 19 Squadron, who came down wounded in the knee on the British side. However, that action was timed at around 13.00, whereas Dannhuber made his claim at 16.20 (G), 15.20 British time.

Nieuport Scouts of 1 Squadron were out patrolling the Oostmeuwerke–Dadizeele–Gheluwe area from 14.50, four machines, led by Captain W T V Rooper, who had scored eight victories since early summer 1917. Flying a Nieuport XXVII (B6767) they were south of Polygon Wood by 15.15, meeting five Albatros Scouts that were chasing two British RE8 observation machines of 4 Squadron. Second Lieutenant H G Reeves attacked one with a black and white fuselage (Jasta 26's markings again) which was right behind one RE. The Albatros burst into flames and crashed near Polygon Wood, the pilot jumping clear of his burning machine while still at 2,500 feet. This was confirmed by 4 Squadron. Reeves then saw an Albatros below him on the tail of Rooper's machine. Reeves dived on it but the German immediately broke off and flew east. Second Lieutenant R A Birkbeck saw Reeves dive on the Albatros but then saw four more Albatoses coming up and went at one which went down with engine on. He fought the others for several moments, one of which began to smoke, then burst into flames and the pilot fell out, or jumped. Second Lieutenant F G Baker scored hits on another and claimed it out of control.

Dannhuber was the only pilot to claim (victory number seven), and perhaps his rapid departure from the fight, with two Nieuports at his heels, helps explain his poor identification. The other problem was that Rooper did not fall inside German lines, but in the trench area, where he tried to make a landing, and broke a leg in the crash. He must have had other injuries too for he did not survive. The Squadron sent out aircraft to locate the crash site and also a breakdown party went forward twice but without success. Rooper was 20, and is buried in Bailleul cemetery.

Jasta 26 lost Leutnant Richard Wagner during this engagement with the RE8s, falling in flames from 800 metres, in Albatros DIII 636/17. His is the only known loss, so perhaps there was some confusion about the number of pilots who fell or jumped from burning aircraft, or was it one and the same man?

Bristol Fighter losses
Considering the extent of the British Front in France, it is amazing how many aces were falling in the more northern part north of Lille, rather than further south around Arras. However, the war had boiled up on the Ypres sector since 4 October as the British began an offensive along a seven-mile front between the Menin Road and the Ypres–Staden railway.

On the 17th another two-seater ace was brought down over Poelcapelle, Captain Arthur Gilbert Vivian Taylor of 20 Squadron. It had been Taylor who had claimed the unidentified scout on 11 September, which this author is convinced sent a shiver through RFC HQ until the time of Georges Guynemer's loss became known. In the event it was an Albatros DV.

On this day, Taylor and his observer, Sergeant W J Benger, were flying a Northern OP (in A7271), taking off at 07.40. The Bristols got into a fight north-east of Zonnebeke and Taylor was shot down. Both men survived the crash but died of their injuries soon afterwards. They were downed by Leutnant Theodor Quandt of Jasta 36, at Poelcapelle, timed at 09.30 (G). Although originally claimed as a DH4, the Germans often confusing BF2b, DH4 and AWFK8s, Jasta 36 claimed two victories, but both were Bristols, the other one having a crewman killed.

Another BF2b crew to fall, this time on 24 October, was that of Captain John Theobald Milne MC, of 48 Squadron, a 22-year-old Londoner (and married), and his observer Lieutenant Stanley Wright MC, from Hull. John Milne had scored nine victories over the summer of 1917 and had become a flight commander. On this day they had taken off at 15.45 on an OP to Dixmude and the Flight had a scrap with German aircraft south of Dixmude at 16.25. Later the French reported seeing a BF2b shot down near Merken. The engagement was with Jasta 29, Leutnant Fritz Kieckhafer claiming a Bristol F2b down over Houthulst Forest at 17.20 (G), and Merken is just west of Houthulst Forest. It was the German's sixth kill, and the Berliner would gain two more before his own loss.

Rhys Davids is killed
The last flurry of activity in October 1917 saw four British aces killed or wounded as the Battle of Passchendaele raged. On the 26th, Canadian Lieutenant E A L Smith of 45 Squadron, who had achieved seven victories, failed to return from a ground strafing sortie which had begun at 08.15 (B5152). He was caught by Leutnant Joachim von Busse of Jasta

3, Emerson Smith being wounded in the left arm and lung, crashed and taken in captivity. On the same patrol, E D Clarke (B2327) was brought down by ground fire, wounded and crashed into the front lines. Edward Clarke had six victories. He received the MC but did not see further combat.

The very next day Arthur Rhys Davids was killed. He was just a month past his 20th birthday, but already he had won the DSO, and MC and Bar, and had 25 victories, having achieved six more after downing Werner Voss in September. The mid-morning patrol for 56 Squadron was along a line Cortemarck–Ruddervoorde–Courtrai–Menin, familiar areas for the SE5 pilots. They took off at 10.35, led by Gerald Bowman, Arthur flying B31. South-west of Roulers they saw two Pfalz Scouts below with six other scouts above them. Too obvious a trap but four SEs dived on the Pfalz, leaving the others to engage the top six. Bowman attacked and followed one Pfalz down but could not finish it off. Climbing back, he found the other action had ended, and on re-forming his pilots, he saw that Rhys Davids was not among them. He had last been seen diving after an Albatros. A further fight ensued before the SE pilots headed home. Three were missing but two others were down safely, but of Rhys Davids there was nothing but an empty sky.

Despite a suggestion that he had been seen attacking a two-seater, it seems he fell to the guns of Leutnant Karl Gallwitz of Jasta 2, his victory over an SE5a being timed at 12.10 (G), 11.10 British time, at Polterrjebrug, about a mile north-west of Dadizeele. He was buried where he fell, and like Voss earlier his final resting place was lost in the turmoil of war.

On the last day of October Sergeant T F Stephenson DCM, of 11 Squadron, was part of a Distant OP over the British 3rd Army Front in A7235, with 1AM S H Platel in the rear seat of his Bristol. Thomas Stephenson had three victories but claimed two more on this patrol. However, in the fighting he was shot up during a fight with Jasta 12 but managed to crash-land in the front line wire. He was okay but Platel had been wounded in the foot and the machine smashed. Another BF2b was shot down by Leutnant Viktor Schobinger of Jasta 12, who claimed at 17.10 (G), south of Marquion, west of Cambrai. Less than a month later, Stephenson was brought down by a direct hit from German AA fire, on the morning of 20 November. He was killed but his Canadian observer, Lieutenant T W Morse, in A7292, was taken prisoner. The 23-year-old from Peterborough has no known grave.

CHAPTER SEVEN

November 1917 – March 1918

The first ace to fall as the battles on the Ypres front died away in lieu of the new offensive down at Cambrai planned for later this month, was a Bristol Fighter man, Lieutenant W G Meggitt MC of 22 Squadron. He and his observers had scored victories over two aircraft in October which, added to the four he had scored while an observer with 25 Squadron in 1916-17, made this Welshman from Newport, South Wales an experienced aviator.

He had earlier served with the Welsh Regiment, but at 13.30 this November day he and his observer, Captain F A Durrad, in B1125, fell in combat with fighters north of Moorslede. Durrad was killed in the action, but William Meggitt, although wounded, became a prisoner. They had been downed by either Heinrich Bongartz of Jasta 36 or Leutnant Werner Dahm of Jasta 26, each of whom claimed a 22 Squadron BF2b between Moorslede and Ledeghem around 13.00 (G). Only Durrad was killed of the four men, so if Dahm's two victims were killed, then Bongartz probably got Meggitt. It was Bongartz's 21st victory (of 33), and Dahm's second of four.

The Naval pilots of 10 Squadron RNAS had continued their support of the RFC on the Western Front during the summer and autumn of 1917. On 12 November they lost another of their Canadian pilots, Flight Lieutenant George Leonard Trapp, from British Columbia. Flying both Triplanes and Camels he had scored five victories by this date, and on the morning of the 12th, he shot down a German two-seater over Couckelaere to make it six. A further OP was flown in the afternoon, Trapp leading eight machines (in B6341) to patrol the line Slype–Dixmude. Visibility was very bad which helped a two-seater they engaged to escape.

Whilst diving on this C-type, Trapp's Camel was seen to break up and fall in pieces. No other enemy aircraft were seen. Unless his Camel simply broke up through structural failure, perhaps the two-seater gunner hit him. The only claim by a two-seater on the 12th was a Sopwith by Unteroffiziers Kuhlmann and Ritzinger of Schutzstaffel 26, west of Weldhoek (or Velhoek, north of Langemarck?).

George Trapp was one of three brothers and two sisters. All three boys joined the flying services, Stanley and George the RNAS, Donovan the RFC. All three were to die, Stanley with 8 Naval in December 1916, Donovan with 85 Squadron RAF in July 1918. Ray Collishaw visited the family on a leave in Canada, and after the war married one of the sisters.

Who got Hans Adam?
Three days later, the 15th, Camels of 45 Squadron, soon to depart for Italy, flew a patrol which attacked a German Rumpler two-seater, which Second Lieutenant Peter Carpenter shot down north-east of Houthulst Forest. Shortly afterwards, the patrol was attacked by a formation of German fighters, and Second Lieutenant K B Montgomery shot down one in flames, which also broke to pieces as it fell, while another went down in flames to Second Lieutenant E M Hand; a third went down out of control to Captain J Firth. These victories were Ken Montgomery's tenth, Earl Hand's first and John Firth's ninth; Hand would become an ace in Italy. All were timed around 09.25–40, with Carpenter's Rumpler a half an hour earlier. Two German fighter pilots were killed at this time and area: one was Leutnant Richard Runge of Jasta 18 (an eight-victory ace), the other Hans Adam of Jasta 6. The Germans noted Runge's loss to AA fire. However, his Albatros DV (5253/17) crashed by the Staden–Langemarck road at 10.40 (G), 09.40 British time, time and place coinciding with 45's action.

By far the most serious loss to the Germans this day was Hans Adam of Jasta 6. Adam had scored his 21st victory on 6 November and had been leading this Staffel since von Dostler had been killed on 21 August. He had already received high awards, being the recipient of the Bavarian Military Merit Order with Swords (4th Class) and the Knight's Cross of the Military Max-Joseph Order, and with it came the title of Ritter (Knight). His next award must surely have been the coveted Pour le Mérite. However, his death precluded this award; instead he received a posthumous Knight's Cross with Swords of the Hohenzollern House Order (in February 1918).

His end came during a fight with Camels and he is noted as falling at 09.20 (G), north-west of Kortewilde, flying Albatros DV 5222/17. If the time is right – and German – then this was an hour earlier than the 45 Squadron combats. However, 65 Squadron were on a Northern OP and at around 07.30 (08.30 German time) they had a fight with eight Albatros Scouts in the Dadizeele area. Two pilots claimed out of control victories, both being their first claims, and Adam is supposed to have been brought down by a novice pilot. Lieutenant G M Cox (B2411) had fired at close range into an Albatros which went down steeply, then into a gliding angle.

As Cox watched, the Albatros dropped its nose and dived vertically. Lieutenant H L Symons (B2418) saw his bullets entering the fuselage of the Albatros he engaged, right behind the pilot's seat. It then went on to its side and dived vertically before it twisted into a spin completely out of control.

Kortewilde is about ten kilometres south-west of Dadizeele, so it is not out of the question that 65 Squadron were involved, and the time is nearer than 45's combat an hour later. Adam's body is supposed to have been discovered in a thicket, stripped and plundered by German soldiers. This on the face of it does not sound as if it was burnt, if either of the 45 Squadron's 'flamers' were involved. Adam, of course, could have jumped from his burning machine, but then his clothes would surely not have been in any fit state to be taken.

The Camels of 45 began with a long-range encounter with 12 to 15 German fighters at 09.35 but the hostiles did not appear keen to fight. Montgomery suspected a trap and he was right, for seven enemy fighters then came down on them from the sun, led by what was described as a 260 h.p. machine (an Albatros DVa?). Montgomery turned his patrol into them. Earl Hand's victory went down over Poelcapelle, 14 kilometres due north of where Adam fell, some 32 Squadron DH5 pilots seeing it catch fire, while Montgomery's fell at Langemarck, in the same area. He and his adversary came at each other head-on but at 30 yards the German machine burst into flames and went under the Camel, but so close that Montgomery reported feeling the heat of the fire, then saw it begin to disintegrate. Firth's is noted as going down east of Comines, which is south-east of Kortewilde, and one wonders if this location is correct. So, 45 Squadron had the kills but not the time or location, 65 Squadron had the time and a better location, and perhaps their possible (out of control) claims were more likely to have been Adam. Who knows? The only other Camels scoring this day were from 10 Naval, but their combat is timed at 12.45 – way too late.

The 15th of November proved expensive for the German fighter arm, for in addition to the above they also lost Leutnant Hans Hoyer, of Jasta 36, who had eight victories. Hoyer was 27 years old and had been in service since 1911, having been decorated while with an artillery unit. Transferring to the air service, and following a period on two-seaters he had joined Jasta 36 in late July 1917. Jasta 36 were in combat with Spads over Zandvoorde at around 12.15 (G), and these appear to be those of 19 Squadron. This unit had been sent off to deal with enemy aircraft working over the Front, Major A D Carter and Lieutenant E Olivier driving a two-seater down out of control over Zandvoorde at 10.15. At about 10.35 the Squadron's aircraft engaged other aircraft above the

Comines canal but had Lieutenant T Elder-Hearn shot down and wounded. Although B3646 crashed, the pilot survived. Two other Spad casualties were recorded as due to ground fire! Was this Jasta 36's fight? It seems a bit early.

Jasta 36 claimed two Spads but neither were confirmed, one by Hoyer. Leutnant Bongartz had also claimed an RE8 east of Merkem at 11.35, and then been wounded north-east of Zillebeke. Some records show this to be by this or another RE8, others note it was a Camel. Everything is around the general Ypres area but it is difficult to say who was fighting whom. Hoyer went down at 12.30 (11.30 British time) at Trichterfelde, two and a half kilometres to the north-east of Tenbrielen, just after the Spad fight. This may have been the scrap that 1 Squadron's Nieuports had at midday over Gheluwe, which is right by the area in which Hoyer came down.

The Nieuport pilots were about to attack a two-seater when they were in turn attacked by three Albatros Scouts, who were then reinforced by five more. Captain Philip Fullard (B6789) attacked one which went down and crashed near Zandvoorde. Another dived past him and getting on its tail, he fired from close range and it dived for 1,000 feet before its bottom left-hand wing broke off and the machine folded up. Second Lieutenant L Cummings sent another that had been attacking Second Lieutenant Ogden down out of control on its back over Gheluwe.

An odd thing about all this is that in the Nachrichtenblat, Bongartz is noted as having shot down an Albatros DIII. Now the nearest thing to a German DIII was a Nieuport Scout because both types had the sesquiplane wing configuration (the lower wing having a much lesser cord area than the upper wing). Both also had Vee-struts between the wings. 1 Squadron lost Second Lieutenant Ogden (B6815), who was taken prisoner. Was this Bongartz's kill, and did Lumsden Cummings then hit him? And if so, was Hoyer Fullard's victim?

These two kills brought the 20-year-old Fullard's score to an impressive 40; he had already been rewarded with a DSO and the MC and Bar. Two days later he broke his leg during a football match on the aerodrome and did not see further action. Had he not done so he may well have been an even higher scoring fighter ace of WWI.

Another German ace wounded on this day was Viktor Schobinger, who had shot down C D Booker back in August. He was still leader of Jasta 12 and had now achieved a score of eight. On this day he was severely wounded in the right foot during air combat over Pervyse and saw no further action in the war. Was he involved in these combats? Jasta 12 were normally a little further south, but had been operating on the more northern 4th Armee Front since 5 November. Unfortunately there is no recorded time of his wounding.

Top left: Sous Lieutenant Adolphe Pègoud, one of the first aces to die in action, brought down by a two-seater crew on 31 August 1915.

Top right: Max Immelmann fell in action on 18 June 1916, but whether by fire from 25 Squadron or by a defective interrupter gear is still unclear.

Middle right: Lt G R McCubbin of 25 Squadron, who with his observer, Cpl Waller, was credited with downing Immelmann.

Bottom left: Ltn Kurt Wintgens, Jasta 1, shot down by S/Lt Alfred Heurtaux, 25 September 1916.

Bottom right: Ltn Willi Fahlbusch, Kasta 1, brought down with his observer by Capt W D S Sanday of 70 Squadron, 6 September 1916.

Top left: Sous Lieutenant Alfred Heurtaux, of N3, went on to gain 21 victories.

Top right: Captain E L Foote, 60 Squadron, was brought down and wounded by Hans Imelmann of Jasta 2 on 26 October, on the British side of the lines.

Middle right: Oblt Stefan Kirmaier, CO of Jasta 2 was downed by Captain J O Andrews of 24 Squadron, on 22 November 1916.

Bottom left: Adjutant M A Lenoir of N3, shot down 25 October 1916.

Bottom right: Major Lanoe Hawker VC DSO, shot down in combat with Manfred von Richthofen, 23 November 1916.

Top left: Ltn Albert Dossenbach (left) and Oblt Hans Schilling (right) had gained eight victories with FA22 by the time they were brought down on 27 September 1916. Schilling and another pilot were killed on 4 December.

Top right: Ltn Gustav Leffers of Jasta 1, shot down by Captain J B Quested and Ltn H J Dicksee of 11 Squadron, 27 December 1916.

Middle right: Captain J B Quested MC, 11 Squadron, as a young lieutenant in the Army Service Corps, 1915.

Bottom left: Ltn Hans Imelmann, Jasta 2, shot down by 4 Squadron, 23 January 1917.

Bottom right: Ltn Hans von Keudell, CO Jasta 27 shot down 15 February 1917.

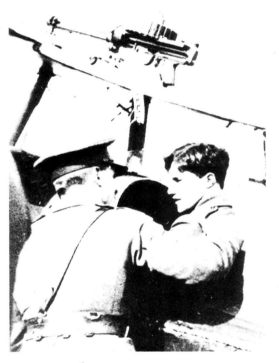

Top left: Lt M Mare-Montembault, 32 Squadron (2nd from right) was taken prisoner after a combat with Adolf von Tutschek, 6 March 1917. Extreme left is Captain W G S Curphy MC.

Top right: Capitaine René Doumer of N76, shot down by Erich Hahn of Jasta 19, 26 April 1917.

Above left: Captain Albert Ball, 56 Squadron, crashed and died on 7 May 1917. Despite being credited to Lothar von Richthofen, Ball's death was an unfortunate accident.

Above right: OfStv Edmund Nathanael of Jasta 5 had scored 15 victories but was shot down on 11 May 1917 by Captain C K Cochrane-Patrick of 23 Squadron.

Right: Jack Malone, 3 Naval Squadron, shot down by Paul Billik of Jasta 12, 30 April 1917.

Top left: Captain W G S Curphey MC & Bar, 32 Squadron (left), brought down on 14 May 1917 by Franz Walz of Jasta 2. The other pilot is Lt L P Aizlewood MC.

Top right: Lt René Dorme, N3, shot down by Heinrich Kroll of Jasta 9, 25 May 1917.

Middle left: Lt L F Jenkin MC, 1 Squadron, was brought down but survived during a combat with MFJ I on 3 June 1917, but was killed in action on 11 September by Otto Schmidt of Jasta 29.

Middle right: Ltn Karl Schäfer, CO Jasta 28, killed in combat by 20 Squadron, 5 June 1917.

Left: Karl Allmenröder, Jasta 11, brought down on 27 June 1917.

Top left: Lt R B Hay MC, 48 Squadron, died of wounds after a combat with Ltn Hans Werner of Jasta 17, 17 July 1917.

Top right: Captain W A Bond DSO MC, killed in action 22 July by AA fire.

Middle left: Walter Göttsch of Jasta 8 had almost a private war with the FEs of 20 Squadron, and they very nearly got him on 29 June 1917; and wounded him on 25 September.

Middle right: Flight Commander J E Sharman DSC, 10 Naval, brought down by AA fire, 22 July 1917.

Right: Captain G H Cock MC, 45 Squadron, in a prison camp show. The 'lady' is Captain D P Collis, 23 Squadron, POW 10 August 1917.

Top left: Oblt Willi Reinhard, Jasta 11, who downed Captain Geoffrey Cock, and went on to command JGI after von Richthofen was killed.

Top right: Adolf von Tutschek was badly wounded by Captain C D Booker of 8 Naval Squadron, 11 August 1917.

Middle left: Captain P B Prothero, 56 Squadron, killed in a fight with Jasta 27, 26 July 1917.

Middle right: Booker, 8 Naval, forced down by Viktor Schobinger, Jasta 12 after wounding von Tutschek.

Left: Viktor Schobinger, Jasta 12, saved his leader by driving down Booker on 11 August.

Top left: Joachim von Bertrab, Jasta 30, brought down and taken prisoner after combat with Edward Mannock of 40 Squadron, 12 August 1917.

Top right: Oblt Edward von Dostler, CO Jasta 6, brought down by 7 Squadron, 21 August 1917.

Bottom left: Oblt Otto Hartmann, CO of Jasta 28, killed in action with 48 Squadron, 3 September 1917.

Bottom right: Captain Clive Collett, 70 Squadron, wounded by Ludwig Hanstein of Jasta 35b, 9 September 1917.

Top left: Oblt Erich Hahn, Jasta 19, had shot down René Doumer on 26 April 1917 but was himself killed on 4 September, by Georges Madon of Spa38.

Top right: Georges Felix Madon, Spa38.

Middle left: Georges Matton, N48, killed in action 10 September 1917, shot down by Josef Jacobs of Jasta 7.

Above: The great Georges Guynemer (inset), Spa3, missing 11 September 1917. One of his Spad fighters on public display after his death.

Left: Josef Jacobs, Jasta 7; he would end the war with 47 victories.

Top left: Kurt Wissemann of Jasta 3 was credited with bringing down Guynemer; he was killed 17 days later.

Top centre: Julius Schmidt of Jasta 6, nearly got A P F Rhys Davids of 56 Squadron on 14 September. Nine days later, RD shot down Voss.

Top right: Kurt Wolff, CO Jasta 29, shot down by Norman MacGregor of 10 Naval Squadron, 15 September 1917.

Above: Werner Voss and his distinctive Fokker Triplane, brought down by Arthur Rhys Davids, 56 Squadron, 23 September 1917. Voss brought down two aces in August and September 1917, Noel Webb of 70 Squadron on 16 August, and Oliver McMaking of 45 Squadron on 11 September.

Middle right: Captain R L Chidlaw-Roberts MC almost didn't reach acedom, so very nearly shot down by Voss on 23 September.

Right: Oblt Hans Waldhausen, Jasta 37, captured after a fight with Booker of 8 Naval and Tudhope of 40 Squadron, 27 September 1917.

Top: Waldhausen's Albatros DV, D2284/17 which became G.75 of British captured aircraft.

Above left: Lt R H Sloley, 56 Squadron, shot down by Xaver Dannhuber of Jasta 26, 1 October 1917.

Above centre: A P F Rhys Davids, having brought down Voss on 23 September, was himself shot down on 27 October, by Karl Gallwitz of Jasta 2.

Above right: Hans Adam, Jasta 6, shot down by 29 Squadron, 15 November 1917.

Left: Ltn Karl Gallwitz; Rhys Davids was his fifth victory.

Top left: Ltn Richard Runge of Jasta 18, shot down by 45 Squadron, 15 November 1917.

Top right: Philip Fullard, 1 Squadron, shot down Jasta 36's Hans Hoyer on 15 November. He is far left in this picture, with H G Reeves, W W Rogers, J B Maudsley, H S Preston, T T Gibbon (RO); C S I Lavers sits on Lt McLaren's shoulders. Fullard, Reeves, Rogers and Lavers were all aces.

Middle left: Ltn Walter von Bülow shot down Lt H G E Luchford of 20 Squadron on 2 December 1917 but on 6 January 1918, as CO of Jasta 36, he was brought down by Willy Fry of 23 Squadron.

Middle right: William Fry MC, 23 Squadron.

Right: Ltn Max Müller, Jasta 28, killed in action with an RE8 crew from 21 Squadron, 9 January 1918.

Top left: Captain G F W Zimmer, the RE8 pilot of 21 Squadron.

Top centre: Lieutenant H A Somerville MC, Zimmer's observer, whose fire set Müller's Albatros alight.

Top right: Lt Ray `Snipe' Foley, observer to Lt Perkins, 27 Squadron; downed OfStv Willi Kampe of Jasta 27, 8 March 1918.

Bottom left: Lothar von Richthofen was lucky to survive encounters with 62 and 73 Squadrons on 13 March 1918.

Bottom right: Von Tutschek, Leader of JGIII, was brought down in a fight with 24 Squadron on 15 March 1918, and found dead by his aircraft.

Top: Captain H B Redler MC (left) while with 40 Squadron in 1917. It was when he was with 24 Squadron in 1918 that he probably hit von Tutschek. Others are: Lt R N Hall, C L Bath and Captain W A Bond DSO MC.

Above left: Lt J A McCudden MC, 84 Squadron, taken prisoner 18 March 1918, following a fight with Hans-Joachim Wolff of Jasta 11.

Above centre: Ltn Hans-Joachim Wolff, Jasta 11. McCudden was his first of ten victories.

Above right: Lt Jean Chaput of Spa57, shot down Ltn Erich Thomas of Jasta 22 on 23 March 1918.

Right: Ludwig Hanstein, Jasta 35b, killed in action with 11 Squadron, 21 March 1918.

Top left: Ltn Erich Thomas, Jasta 22, POW 23 March 1918.

Top right: Captain J L Trollope MC, 43 Squadron, shot down 28 March, taken prisoner and lost an arm.

Middle left: Ltn Hans Müller, Jasta 18, would score 12 kills, but was lucky not to have been killed by victory number three, 27 March 1918.

Above: Captain G B Moore, 1 Squadron. His SE5 was blown apart by an AA shell on 7 April 1918.

Bottom left: Captain T S Sharpe DFC, 73 Squadron, brought down by Manfred von Richthofen, 27 March 1918.

Top left: Manfred von Richthofen, fell over Morlancourt Ridge, 21 April 1918.

Top centre: Heinrich Bongartz, Jasta 36. Twice the British nearly got him, on 13 July 1917 and 29 April 1918.

Top right: Ltn Hans Weiss, Jasta 11, shot down by 209 Squadron, 2 May 1918.

Bottom left: Captain T Durrant, 56 Squadron, shot down by Hans Kirschstein of Jasta 6, 16 May 1918.

Bottom right: Major Raoul Lufbery, hero of the Lafayette, killed commanding the 94th Aero, 19 May 1918, shot down by a two-seater.

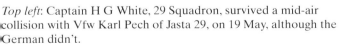

Top left: Captain H G White, 29 Squadron, survived a mid-air collision with Vfw Karl Pech of Jasta 29, on 19 May, although the German didn't.

Top right: Lieutenant Paul Baer of the 103rd US Aero, brought down and taken prisoner after a fight with Jasta 18, 22 May 1918.

Above left: Ltn Walter Böning, Jasta 76.

Middle right: What happened to Rudolf Windisch of Jasta 66, after falling into French lines on 27 May 1918, still remains a mystery.

Right: Captain John Todd MC, 70 Squadron, was in the fight in which Ltn Walter Böning was nearly killed, 31 May 1918.

Top: Major R S Dallas, CO of 40 Squadron, was surprised by Ltn Hans Werner, CO of Jasta 14, 1 June 1918. He is seated in the SE5 in which he fell.

Above left: Ltn Kurt Wüsthoff, shot down and captured, 17 June 1918.

Above right: Fritz Loerzer of Jasta 63, taken prisoner 12 June 1918.

Right: Major J C Callaghan MC, CO 87 Squadron, shot down by Franz Büchner, leader of Jasta 13, 2 July 1918.

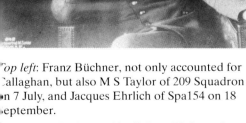

Top left: Franz Büchner, not only accounted for Callaghan, but also M S Taylor of 209 Squadron on 7 July, and Jacques Ehrlich of Spa154 on 18 September.

Top centre: Carl Menckhoff, Jasta 72, brought down and captured on 25 July.

Top right: Lt W O Boger, 56 Squadron, killed in action 10 August.

Middle left: A trio of aces of JGII, Paul Strähle, Jasta 18, Josef Veltjens, Jasta 18, and the great Rudolf Berthold, who commanded JGII. Veltjens may have downed Boger on 10 August, while Berthold was badly wounded on this same day.

Above: Lt Walter L Avery, 95th Aero, the embryo pilot who downed Carl Menckhoff.

Bottom left: Erich Löwenhardt, leader of Jasta 10, another victim of a mid-air collision, 10 August 1918, after gaining 54 victories.

Top left: Ltn Pippart of Jasta 19, baled out on 11 August 1918 but his parachute failed to open.

Top centre: Major C D Booker, seen here with A M Shook (left) (the day they both received the Croix de Guerre), was killed on 13 August 1918 in combat with Ulrich Neckel, Jasta 12.

Top right: The mercurial Louis Bennett Jr, 40 Squadron, downed by ground fire on 24 August 1918.

Above: Ulrich Neckel, Jasta 12.

Right: Captain J K Summers MC, shot down and captured by Lothar von Richthofen, 12 August 1918.

Top left: Captain A A N D Pentland MC DFC, wounded 25 August 1918 while B Flight Commander, 87 Squadron.

Top right: Rudolf Klimke was one of the Jasta 27 pilots in action with the 17th Aero on 26 August. He was wounded on 21 September 1918.

Middle left: Lt R M Todd, 17th Aero, taken prisoner following a fight with JGIII, 26 August 1918.

Above: Captain H A Patey DFC, as a POW; downed by Lutz Beckmann of Jasta 56, 5 September 1918.

Left: Lt J O Donaldson, 32 Squadron, POW 1 September after meeting Ltn Theodor Quandt of Jasta 36.

Top left: Lt David Putnam of the 139th Aero, shot down by Jasta 15's Georg von Hantelmann, 12 September 1918.

Top right: Hantelmann also brought down Lt M J P Boyau of Spa77 on 16 September 1918.

Bottom left: Georg von Hantelmann, Jasta 15.

Bottom right: Joe Wehner of the 27th Aero, also fell to von Hantelmann, 18 September 1918.

Top left: Jacques Ehrlich (right) of Spa154, captured after being shot down by Franz Büchner, 18 September. At left is another ace, Paul Barbreau, also of Spa154. Sgt Abbott centre.

Top right: Ltn Fritz Rumey of Jasta 5 had downed several aces but fell himself on 27 September.

Middle left: Ltn Frank Luke Jr, 27th Aero, died of wounds received during a balloon attack, 29 September 1918.

Bottom left: Ltn Fritz Höhn, Jasta 41, shot down by Adjutant Le Petit of Spa67, 4 October 1918.

Bottom right: Lt W W White, 147th Aero, died ramming Wilhelm Kohlbach of Jasta 10, 10 October 1918.

Top left: Howard C Knotts, 17th Aero. Taken prisoner on 14 October, he destroyed several Fokkers by sabotaging a train on his way to prison camp.

Top right: Captain W G Barker won the VC for an action on 27 October 1918 although the unit he was fighting remains a mystery.

Middle left: Captain G W Wareing, 29 Squadron, killed by Josef Raesch of Jasta 43, 27 October 1918.

Bottom left: Captain A A Callender, 32 Squadron shot down by Jasta 2, 30 October 1918.

Bottom centre: Lt R DeB Vernam, 22nd Aero, died of wounds after being brought down on 30 October 1918.

Bottom right: Lt Jim Beane, also of the 22nd Aero, killed 30 October 1918.

Ground fire brought down another ace on 22 November, the Scotsman Captain Gerard Bruce Crole MC of 43 Squadron. An Oxford BA, Crole had scored five victories with 40 Squadron in the summer of 1917 before becoming a flight commander with 43 Squadron. Flying Camel B6267 he took off on a Special Low Recce sortie at 07.55 to Douai and did not return until 14 December 1918 – from prison camp. He and another pilot, Lieutenant W E Nicholson (B2351) fired into a company of German infantry on the Douai–Cambrai road and then Crole went missing. On his return from Germany he confirmed his petrol tank had been shot through whilst ground strafing, and he had been wounded, forcing a crash-landing.

A two-seater gets Böhme
Leutnant Erwin Böhme was one of the unluckiest pilots in that on 28 October 1916 he collided with his friend and Staffelführer, the great Oswald Boelcke, which resulted in the latter's death. Despite this, Böhme had himself become a high scoring fighter pilot, and by November 1917 was himself a Staffel leader, of Jasta 2, had scored more than 20 victories and been rewarded with the Pour le Mérite on the 24th of that month. Five days later he was killed.

His 24th and last victory fell to him at 12.55 (G) on the day of his death, a Sopwith Camel over Zonnebeke. Later that afternoon he was in the air again. So too was a 10 Squadron AWFK8 'Big Ack' B324, crewed by Second Lieutenants J A Pattern and P W Leycester, having taken off at 14.45 to take panoramic photos of Polderhoek Chat and Becelaere. They had exposed 14 photographic plates but were then interrupted by the arrival of three German fighters that swooped down on them from clouds at 14,000 feet above Zonnebeke. Leyster immediately grabbed his Lewis gun and began firing. After just a short burst the leading Albatros erupted into flames, dived earthwards and crashed. The other two fighters sheered off and followed the burning machine down but they could only watch as their leader dived into the ground.

The very next day, Julius Buckler was forced to land. Flying with Jasta 17 he had been in combat with an RE8 of 9 Squadron at around 13.15, which came down in the front lines. The crew escaped but the two-seater was shelled into fragments. Little is known of why Buckler came down, unless the RE8 crew managed to hit him in the same action. Buckler had 29 victories, and scored again by the end of the month to receive the Blue Max. His final score was 36; not even a wound in May stopped his run, although it put him in hospital for eight weeks.

Luchford's end

Harry Luchford had now scored 24 victories flying both FE2d and BF2b two-seaters. In fact he was among the highest scoring of the two-seater teams. This former bank clerk from Bromley in Kent was only 23 but had the MC and Bar to show for his prowess. On 2 December he flew a patrol, for once without his usual observer, Lieutenant V R S White MC, but with Captain J E Johnston (A7292). He had not scored since October so perhaps he'd been on leave. If so it was always a dangerous time returning to the war after a period away from the action. Pilots built up a sort of awareness of combat and no matter how good one was, it needed time to get back into the swing of things once one returned to active duty.

Luchford and Johnston were last seen in action over Passchendaele at around 10.30 and came up against the experienced Leutnant Walter von Bülow-Bothkamp, also aged 23, who had achieved 27 victories to date. He had received the Pour le Mérite, Knight's Cross of the Hohenzollern House Order and the Saxon Military St Henry Order. He had been leading Jasta 36 for some months after service as a two-seater pilot in both France and the Middle East, returning to fly with Jasta 18 in early 1917. Luchford went down at 11.35 (G) to crash near Becelaere. He has no known grave, but Johnston survived as a prisoner. Not long after his 28th victory, von Bülow was given command of Jasta 2 'Boelcke', following Böhme's loss.

Ground fire caused the loss of a 56 Squadron ace on 19 December, Captain Richard Maybery MC and Bar, flying SE5a B506. He took off at 12.20 and was last seen going down, possibly out of control, over Bourlon Wood after bringing down an enemy aircraft in flames at 13.00. Despite the combat, it is understood that Maybery was shot down by ground fire, claimed by K-Batterie Nr.108, and was buried by the battery commander where he fell. Maybery, a Welshman and former Lancer, had 21 victories.

Hess goes down on the French front

Another veteran of long experience was Leutnant Ernst Hess. A pre-war pilot, he had flown two-seaters, Fokker monoplanes, then flown with Jasta 10, Jasta 28w and now commanded Jasta 19. He had brought his score to 17 by the end of October. There are two suggestions of how he met his end. Adjutant De Kergolay of N96 claimed an Albatros Scout over Fresnes at 13.10, his first victory. Hess is also supposed to have gone down in flames during a combat with a Letord three-seater east of Fresnes–Modelin Farm Road. French records seem to indicate a combat with a Caudron R4 of Escadrille C39, in which the observer, Sous

Lieutenant Dutrou, and the gunner, Soldat Tissier, were both wounded before the German fighter went down. Hess had been flying Albatros DVa 5347/17.

Another ace to fall in December was Ontario-born Alfred Edwin McKay MC, who had scored ten victories flying with 24 and 23 Squadrons. It was the day after his 25th birthday, 28 December, that he was killed.

He had led his Flight on an OP to Checkern–Quesnoy (B6734), taking off at 11.45 hrs. He was last seen at about 12.15 between Gheluvelt and Dadizeele, diving on a German two-seater and then spinning. He was the victim of either Schutzstaffel 26 or 28, both of which had combats this day. He has no known grave.

1918 – Jasta 'Boelcke's' new commander falls
Walter von Bülow had not scored since he downed Harry Luchford on 2 December. Since leaving Jasta 36 to command Jasta Boelcke, the winter weather as much as anything else had curtailed operations. However, he was airborne on the afternoon of 6 January 1918 to become the first ace to fall in the New Year.

Flying a Reserve Patrol over the Passchendaele area at 14.00 hrs was Captain William Mays Fry MC, in Spad B3640 of 23 Squadron. Willie Fry was a very experienced fighter pilot, having cut his combat teeth with 60 Squadron during 1917, after serving with 12 and 11 Squadron on two-seaters in 1916. Now, the 21-year-old was a flight commander after helping to form 87 Squadron in England. To date he had eight victories.

Fry spotted around five Albatros Scouts flying west and fighting with a group of Camels. He dived on one, firing a short burst of about 20 rounds from right behind. The Albatros rolled over and went down in a steep spiral. Willie watched it go down and crash in the shelled area south of Passchendaele, either on the British side or at least in the trench area.

The Camels were from 70 Squadron and two of its pilots also claimed they engaged the Albatros. Captain Kemsley was leading the patrol and he and Lieutenant F Quigley (B2447) engaged the German patrol. Quigley shot down one which he saw crash near Stadenberg. Then he saw another on the tail of a Camel and attacked this machine, as did Kemsley. Second Lieutenant F C Gorringe (B6426) also came in for an attack, saw it go down, roll on to its back and crash into the ground near the front lines. Both Francis Quigley, a Canadian, and Frank Gorringe were aces, the former with ten victories, the latter with nine. Both men and Fry shared in the victory, so by the end of this day, Quigley had raised his score to 12, Gorringe to ten and Fry to nine. Quigley, 23, from Toronto, would run up a score of 33 victories by March but was then wounded. He received the MC and Bar but died in the great influenza epidemic in October 1918. Gorringe, from Eastbourne, was 28 and by the end of the war had the MC and DFC plus 14 victories. Fry ended his war with 11. He died in 1992.

Von Bülow's replacement is lost too
Acting command of Jasta Boelcke passed to Leutnant Max Müller, a 31-year-old Bavarian. Whether he would have been given full command of the Jasta is doubtful due to the strict officer corps within the German services hierarchy, and Max Müller had been an NCO for a long period. Only his achievements in the air had given him officer status, as late as August 1916. It had been a field promotion into the regular forces rather than the reserve, but the higher echelons were still sticklers for protocol. Müller nonetheless took over on 6 January after von Bülow's loss, and lasted just three days. By that date he had achieved 36 victories, with at least two more unconfirmed, and he held the Pour le Mérite and other high awards. The story goes that the previous day he had been lecturing pilots on how to attack the British RE8 machine, and as luck would have it, on 9 January he was on patrol behind the front and spotted an RE8. Not that he had a vast amount of experience with the RE8 type; in his score he had only two successes against them, the second as far back as June 1917.

A 21 Squadron RE8 (B5045), crewed by Captain G F W Zimmer and Second Lieutenant H A Somerville MC, took off on a photo Op at 10.45 to the Moorslede area. While engaged in this duty, photographing the Corps Counter Battery area near Moorslede, they were attacked by seven Albatros Scouts of Jasta 2, who approached from the north at 11.35. One of them got to within 25 yards of the RE's tail and Henry Somerville, who had won his Military Cross with the Royal Sussex Regiment in late 1917, fired with his rear gun into the fighter. After 50 rounds he saw it sheer away and burst into flames, then dive out of control between Moorslede and Dadizeele (DVa 5405/17). A few minutes later, SE5s of 60 Squadron came upon the scene and attacked the other Albatros Scouts without result, although with the report of one Albatros crashing in flames, two pilots received credit, despite not initially observing results of their attack. 60 Squadron's version was that '. . . SE5s fired into three of the enemy aircraft at very close range but were not able to observe results. All EA dived east.' Captains F O Soden and R C Chidlaw-Roberts were given half share in one victory following an 11 Wing report of a machine of unknown nationality falling in flames at 11.55. However, there is no doubt that Zimmer and Somerville downed Müller.

George Frederick William Zimmer, from Penge, London, was 28 and had been in France since July 1917, firstly with 21 Squadron, then briefly with 10 Squadron, then back to 21 as a flight commander, arriving the previous day, the 8th. He later received the DFC and died in March 1971. His observer on this day did not survive that long. He moved to 82 Squadron soon afterwards to operate on AWFK8s. He was shot down and killed on 28 March 1918, and is buried at Dompierre.

Long on experience, short on luck
There were few British squadron commanders who were allowed to fly operationally as the loss of such experienced men would not help the cause. From time to time, there were definite orders about such men absolutely not flying. There were some of course, who were more commanders than fliers and were quite happy to let their flight commanders carry out the battle orders. However, there were still a number of men who were not content to sit and watch, but flew, orders or no orders.

One such man was Major F J Powell MC, CO of 41 Squadron. Powell had joined the RFC from the Manchester Regiment and had obtained his pilot certificate in March 1915. He had flown long and hard with 5 Squadron with both the Vickers FB5 'Gunbus' and one of the first pusher FE8s. He had won his MC in January 1916 and became a flight commander with 40 Squadron later in the year, having scored a number of successes with 5 Squadron. Due to the type of scoring claims in those early days, his tally looked something like one and one shared destroyed, one 'out of control' plus three or four more driven down. He was made commander of 41 Squadron on 2 August.

On the afternoon of 2 February 1918 he took off in SE5a B8273 and over Auberchicourt, to the east of Cambrai, ran into Leutnant Max Kuhn of Jasta 10. Kuhn had been with the Jasta since July 1917 and is supposed to have scored two victories with an unknown unit before this. He had scored nothing with Jasta 10 but got lucky on this day. Five Albatros Scouts came down on Powell and a burst from Kuhn hit the Major in the arm and damaged his engine. With ebbing strength, Powell got his machine down, actually finding an airfield on which to force-land, but he was quickly surrounded and quite unable to set fire to his machine. Kuhn had got his one and only victory with Jasta 10 that had put the vastly experienced Powell behind wire for the duration. He was repatriated in December.

Black days for the Navy
February 1918 proved a bad month for the RNAS flyers. On the 3rd, 9 Naval Squadron lost Flight Commander R R Winter. A Londoner, Rupert Winter had flown with 6 Naval until it disbanded, whence he had

joined 9 Naval. By the end of 1917 he had four victories; knowledge of his fifth was to be short-lived.

On the afternoon of 3 February, 9 Naval flew a patrol and got into a fight with Fokker Triplanes of Jasta 26 near Roulers, joined by Camels of 54 and 65 Squadron. Despite the Germans claiming four victories, only one was actually lost, Winter, while a pilot of 65 Squadron got down with a damaged machine and a 54 Squadron pilot returned wounded. It was unusual for so many credits to be given by the Germans, but for once they overclaimed considerably. The claimants were Offizierstellvertreter Otto Esswein three, Vizefeldwebel Otto Fruhner one. Winter was seen to shoot down one Triplane in company with FSL M A Harker (B3781), but moments later his B6430 went down and its wings folded up. Jasta 26 suffered no pilot losses.

Two days later 8 Naval Squadron lost Flight Lieutenant Harold Day DSC, a veteran air fighter who downed his 12th opponent in this his last fight. The Camel Squadron had been on patrol from around midday and at 12.45 had been in action with some Albatros Scouts south of Pont à Vendin. 8 Naval were operating much further south at this time, supporting the RFC on the Lens front. Day was reported to have shot down one Albatros and was diving upon another when his Camel (N6379) broke up. However, he appears to have been claimed by Leutnant Gunther Schuster of Jasta 29, who claimed a Camel north-west of Annay at 13.45 (G). As Day crashed at 12.45 and Pont à Vendin is north-west of Annay, this fits the loss. Schuster thus achieved his fourth victory of an eventual six.

On 18 February, 8 Naval lost another veteran pilot, Flight Commander G W Price DSC and Bar. Guy Price, an Irishman, had earlier flown with 13 Naval and sported a 'Captain Kettle' type beard. Since December 1917 he had achieved a dozen victories, but on the 18th, while flying a ground strafing sortie, he was caught low down by a pilot from Jasta 23. Theodor Rumpel made his claim at 13.20 (G), at Givenchy, and Price had been flying towards the Vitry–Douai area, and had gone down at 12.20, so both fit. It was Rumpel's fifth and final victory; he was severely wounded in action on 24 March. One disadvantage Price had was that he was flying a brand new Camel, B7204, which had only been received the previous day from the Naval Air Park, so it was unfamiliar to him.

Hans Klein is wounded

Oberleutnant Hans Klein had scored 22 official victories by the end of November 1917, and had received the coveted Ordre Pour le Mérite at the beginning of December. He was among the top scorers in Jasta 4 of the Richthofen Jagdgeschwader, and had been given command of JGI's Jasta 10 in September, after being wounded the previous 13 July in a fight with 29 Squadron's Nieuports (probably Lt H M Ferreira) when his score was 16. He had returned to the front and Jasta 10 at the end of September and added six kills during the early winter period.

On 19 February 1918, Klein was trying to get back into the scoring habit again not having found his form since the previous year. Flying a Pfalz DIII (4283/17) he and his men engaged a formation of Camels of 54 Squadron just before 13.00 (G) in the Monceau area. Lieutenant N Clark (B5421) and Second Lieutenant A T W Lindsay (B9281) attacked five German single-seaters escorting a two-seater. In the fight that developed, both men fired into one fighter which began to fall vertically until they saw it hit the ground. How much of a crash is not determined, but Klein had been badly wounded in the right hand, losing the thumb. He did not return to operational flying, although he did serve as a ground officer with his Jasta 10 under Erich Löwenhardt. Norman Clark was killed in action on 18 March, while Lindsay was captured on 26 March.

Lost at sea

Flight Commander M J G Day DSC had been retained as a test pilot on the Isle of Grain but eventually managed to get to the war, joining 13 Naval Squadron (formally the Seaplane Defence Flight) at St Pol, just behind the harbour at Dunkirk. Within the first few weeks of 1918 he had claimed five victories and won the DSC, whilst operating along the North Sea coast, but his luck ran out on 27 February. Flying on patrol this day the Camels were several miles out to sea after reported German seaplanes. They spotted six seaplanes at about 13.00 and made an attack but one, crewed by Flugmeister Dreyer and Leutnant Frantz of Seaflug II, got in a telling burst on Miles Day's Camel (N6363, although the Germans thought it was an English Spad) and set it on fire. Day got down on the sea rapidly, although his machine broke up. He was seen clinging to the wreckage, but despite later searches by aircraft and destroyers, he was not found.

Bombers get Willi Kampe

Offizierstellvertreter Willi Kampe had scored seven victories between 15 February 1917 and 4 January 1918, flying with Jasta 27. In our book *Above the Lines* my co-authors and I chose 19 Squadron's Sopwith Dolphins as being the victors over Kampe, but it seems now that they shot down Leutnant Heinrich Mindner of Jasta 51. We were somewhat confused by a claim for his eighth victory on the day of his death, but further investigation shows this to have been a British two-seater, not a Dolphin, although the time and area were not dissimilar. It now appears that Kampe was attacking a DH4 of 27 Squadron, crewed by Lieutenants

Jukes F R I Perkins and Raymond G 'Snipe' Foley MC (B2904), who were both Canadians. Foley got his nickname when with the 2nd Battalion, CEF, because he was a fine shot.

Six DH4s were bombing Busigny railway station and were attacked as they turned for home just after midday. The fighters managed to get round behind the formation and a fight started over Walincourt. Three Albatros Scouts were seen to spin away after meeting return fire from the bombers, one being seen to go down very positively out of control after closing in on Foley and Perkins. Just moments later their 'Four' dropped out and went down burning. There was some heavy AA fire during this action, but Jasta 27 credited Kampe with his eighth kill, although it had cost him his own life.

Lieutenants James Gray and James A McGinnis (B2088) made out this combat report of the action:

> 'The formation of four DH4s dropped bombs on objective (Busigny). When turning we were attacked by 12 EA Albatros Scouts. One of our machines, Lt Perkins, turned to engage three EA with his forward gun but was seen to go down in flames. Before he went down, Lt Foley, his observer, appeared to hit one of the EA which went down in a spin, apparently out of control.
>
> Lt McGinnis put a long burst into another EA which stalled, turned over on its back and went down. We were forced to land north-east of Peronne with engine trouble.'

In addition, the crew of Lieutenants A H Hill and K G Mappin (B2107) also claimed a fighter out of control, and they too were forced to land on the British side with engine trouble, while under attack. Therefore, in reality, any one of the three crews above could have been the ones to hit Kampe, but with Foley's skill with a gun, it was probably him. Perkins and Foley were on their tenth mission.

No.3 Naval Squadron gets Buddecke
On 10 March the times changed. Until 15 April both the British and German clocks showed the same times so combats will occur at the same time during the following actions.

Rudolf Berthold had been given command of Jagdgeschwader Nr.II and would take over officially on 10 March, so he handed over command of his Jasta 18 to Oberleutnant Hans Joachim Buddecke (full name Adolf August Hans-Joachim Buddecke); both were holders of the Ordre Pour le Mérite. Buddecke had started his war flying Fokker Eindekkers in France in 1915 with FA23, and had saved his friend Rudolf Berthold, who was being attacked by another aircraft, on 15 September. However, he gained his fame out in the Middle East flying single-seaters with FA6 on the Greek front, where he became known as the Eagle of the Aegean. The Turks called him the Hunting Falcon. He became the third recipient of the 'Blue Max' in April 1916, at which time he had gained seven official victories with at least another six unconfirmed. He returned to France to command Jasta 4 in the late summer and claimed three more certain kills, but then returned to Turkey in mid-December, this time flying with FA5, bringing his score to 12. Early in 1918 he returned to France once more and took command of Jasta 30, shooting down a Camel for his 13th victory on 19 February. He then went to Jasta 18 as deputy commander under his old friend Berthold, the Staffelführer, just as he was starting to form his JGII, Berthold still suffering from his injured and almost useless right arm, which was refusing to heal properly since the previous October.

However, combat had changed a good deal on the Western Front since the calmer days of 1916 and Berthold's experience, while great, was no real match for the aggressive and versatile Allied airmen. Thus on 10 March, a patrol was flown led by the two 'aces', leading another five men of Jasta 18. At 13.15, near Lens, they spotted a formation of Camels and joined combat. Flight Lieutenant Art Whealy, a Canadian, was leading a patrol of 3 Naval Squadron, and dived with his men onto the seven Albatros DVs. He attacked one but then saw another to his left about 500 feet below and heading away. Whealy immediately dived on this one's tail and began firing from 100 yards. The Albatros half turned on to its back and began to go down in a series of stalls and spins. What Whealy hadn't realised was that this second Albatros, flown by Buddecke, had seen the Camel attack Berthold and had tried to intervene, only to shoot past underneath the Camel, coming into view of the Canadian in a position too good to miss. Although two other pilots also claimed 'out of control' victories – Flight Lieutenant A B Ellwood and Flight Sub Lieutenant J S Britnall – credit for Buddecke seems to have gone to Whealy for although he lost his spinning victim in ground haze, he had felt certain he'd hit the pilot and other 3 Naval pilots confirmed seeing it crash into the ground two miles east of Lens, near Harnes. This time, although he had saved Berthold for the second time, it had cost him his life. Buddecke was buried in Berlin on the 22nd.

The aces fall before the Great Offensive
The Germans, with troops released from the Russian Front following the ceasefire there, and wanting to force a decision on the Western Front before the full weight of the American involvement started, planned a huge offensive for March. The German Air Service kept up strong

pressure along the British Front as March progressed in order to keep the Allied airmen's eyes and cameras from discovering the vast movement of troops and supplies in the rear areas. However, with the increased activity came an increased risk of casualties.

Not all German Marine pilots flew with the Marine-Feld Jastas; several were attached to Army Jastas too. One of these, Oberleutnant zur See Konrad Mettlich, had flown Fokker Eindekkers in 1916 with an Armee Staffel. Then, following a brief stint with MFJ I, he went to Jasta 8, taking command of this unit in July 1917. By March 1918 he had scored six victories, but met his match on the 13th.

The SE5s of 84 Squadron were on patrol north-east of St Quentin shortly after midday, five aircraft of C Flight bumping along at 17,000 feet. They observed several Albatros Scouts making life difficult for three Camels, the SE5s falling on the V-strutters from out of the sun. Second Lieutenant P K Hobson (D260) singled out one just a little north-east of the others and closed in as he began diving eastwards. Closing right behind him, Percy Hobson, a 24-year-old from Bungay, Suffolk, opened fire with both guns. His fire smashed into the Albatros, Hobson watching as the German's top and bottom left wings tore away; then the fighter commenced spinning. Hobson then went after two more but they quickly broke away eastwards.

Back at base, Captain F E Brown MC (B8337) confirmed the Albatros going down in pieces, having himself shot down an Albatros which he saw crash into Homblières village. Second Lieutenant C L Stubbs (B5308) sent another down out of control over Chardenvert, but lost sight of it in ground mist at 1,000 feet. Mettlich's wreck crashed near Remaucourt, Hobson having obtained his third of an eventual seven victories. Jasta 8 also had Vizefeldwebel Adolf Besemüller severely wounded in this combat and he died the next day, presumably Frederic Brown's victim. Brown was already an ace and this was his sixth victory. From Quebec, Canada, he would raise his score to ten before the month was out. Jasta 8 had been up against some good pilots, for even Charles Stubbs reached acedom.

Lothar von Richthofen's dice with death
As an historian who has studied the British and Commonwealth air fighters of WWI for many years, I have reluctantly had to face the fact that overclaiming went on at an alarming rate. With only a few exceptions, most victories were claimed in good faith, even if some with a bit of 'tongue in cheek'. What, however, I have never discovered is how and why 'higher authority' condoned these 'claims' which so often went towards awards and decorations. Was it merely to keep up morale?

A case in point is the events of 13 March 1918, during which Lothar von Richthofen went down and so very nearly was killed. Lothar was rather more reckless than his astute elder brother Manfred, but would nevertheless achieve 40 victories in the war. By this March day he had scored 29, had commanded Jasta 11 since November 1917 and been awarded the Blue Max. In a fight with Bristol Fighters of 62 Squadron and Camels of 73 Squadron, JGI's Triplanes had arrived on the scene just as the Bristols, who had been stalking another formation of German fighters for some time, had decided to go home. Inexplicably, as the Bristol's leader turned, a couple of his men suddenly dived on the lower formation, obviously oblivious of the approaching Triplanes, which the Bristol leader had seen. This forced the Bristol leader to go to their aid and a fight started.

It was 10.15, east of Cambrai, and the Camels joined in. Captain A H Orlebar (B7282), who in later years would head the British Schneider Trophy team, sent an Albatros down out of control, then attacked a Triplane, which nose-dived and shed its top wing.

In the same fight, the Bristol leader, Captain G F Hughes with Captain H Claye in the back (C4630), saw a Triplane come down on them from behind. Hugh Claye opened fire and saw the top wing of the Triplane disintegrate and the fighter fell away out of control. Hughes and Claye were then attacked by Lothar's big brother and were very nearly shot to pieces, but survived.

Lothar survived a crash-landing but was badly injured, not returning to his command until July. He was still in a hospital bed when the news of Manfred's death reached him. The question, of course, is how did RAF HQ allow credit for this Triplane to go to Orlebar *and* the Hughes/Claye team? Perhaps it didn't matter. At least the German system of accrediting victories to just one man, sometimes after arbitration, kept individual victory scores to a minimum, although how they might have assessed this victory is difficult to imagine. Hughes and Claye received MCs soon afterwards, but Orlebar, who ended the war with seven victories was never rewarded.

Who downed von Tutschek?

Two days later another German 'Kanone'* – went down, although for some time it was not certain how he had met his end, or by whom. Rittmeister Adolf Ritter von Tutschek was a Bavarian and holder of the Pour le Mérite. We read earlier how he was wounded back on 11 August 1917. Having now returned to the Front, he was commanding Jagdge-

---

* Kanone; big gun, the German name for an 'ace'.

schwader Nr.II, having raised his score to 27. On this morning, von Tutscheck took off at 10.00 leading a mixed formation of Triplanes and Albatros Scouts.

Shortly after 10.00 on this 15th day of March, 24 Squadron were on patrol over Premonte at 14,000 feet, led by Captain B P G Beanlands MC (C1081), who was shortly to go home on rest. His patrol consisted of some experienced pilots, Lieutenants A K Cowper (C5428), R T Mark (C9494), H B Redler (B79) and H B Richardson (B8257). All, except Redler, were aces by this time, and even he soon would be. There were several enemy machines about, the SE5 pilots counting three Rumpler two-seaters, about ten Albatros Scouts and three Fokker Dr1 Triplanes. There were several combats. Herbert Redler, who had flown with 40 Squadron the previous summer, climbed to 16,000 feet into the sun and saw the three Triplanes, and three Albatros Scouts, going west, 1,000 feet below. He crossed to the east of them and dived on the topmost 'Tripehound', firing 40 rounds into it, hammering away until he all but collided with the dark green coloured machine, with its white engine cowling and wing streamers. From experience he knew he had selected the leader. The Triplane stalled to the right and went down in a spin. Redler turned and watched it spin down for some 2,000 feet above Bancourt before having to retire on account of the remaining Germans turning their anger on him. He did not see the Triplane crash but it was obviously out of control.

There have been conflicting accounts of von Tutschek's death, but the main story noted that the German ace had recovered from the spin and had made a forced landing. His two companions, Oberleutnant Paul Blumenbach (CO Jasta 12) and Leutnant Paul Hoffmann (also Jasta 12), had followed him down and were relieved to see him climb from the cockpit and wave. A short time later, as German troops arrived at the scene, they found the pilot dead on the ground, a glancing head wound the only sign of injury. All believed he had been surprised by the attacking enemy, and they concluded he had been hit a glancing but severe blow to the temple, from which he succumbed soon after waving to his men. He had landed at Bancourt, two to three miles south-east of Premonte, and his well-known Triplane was later smashed by Allied artillery fire. Later the German pilots confirmed that the first they had known of the SE5's attack was when von Tutschek's machine began to spiral down.

Killed over Verdun

The French had been fighting and dying around Verdun for what must have seemed years. It was their open wound on the Western Front. In the air the French airmen fought no less gallantly over this fortress city.

On 16 March, one of their aces fell, Sous Lieutenant François De

Rochechouart de Mortemart, Prince of Tonnay-Charente, of Spa 23. Yet another Paris-born airman, he was six days short of his 37th birthday, so was no impressionable youth. Following service with a cavalry unit at the beginning of the war he had advanced to aviation and his seven air victories had brought him the Légion d'Honneur.

Flying Spad XIII No.2167 he was shot down over Consenvoye in the Verdun sector. There are two main candidates for bringing him down, Leutnant Arno Benzler, CO of Jasta 65, who claimed a Spad at Malancourt at 13.00 for his third victory (of nine); and Leutnant Bleibtreu of Jasta 45, another Spad over Maasbogen at 17.30, his first and only victory. At this stage, the time of the Frenchman's loss is not known.

No Ordre Pour le Mérite for Hans Bethge
As we have seen before, it was not always fighter versus fighter battles that cost the aces their lives, but action against two-seaters as well, and not just the versatile Bristol Fighters. On 17 March, Oberleutnant Hans Bethge of Jasta 30, having achieved his 20th victory a week earlier, was eagerly awaiting the expected accolade of the Ordre Pour le Mérite, usually awarded after this number of kills. Unfortunately such awards were not made posthumously, even if earned.

Half an hour before noon on this day, 57 Squadron's DH4 bombers were on a combined bombing and photo mission north of Menin. Immediately after dropping their bombs on Linselles, the crew of A7901, Captain A Roulstone MC and Second Lieutenant W C Venmore, were attacked by six enemy scouts. Nottingham-born Alex Roulstone was no push-over, having been at the front for several months not only with 57 but earlier with 25 Squadron too. He and his various observers had already notched up more than half a dozen victories over other would-be assassins. Roulstone pulled his bomber up into an Immelmann turn, then, having reversed the tables somewhat on his attackers, came down behind one Albatros firing 50 rounds from 50 yards, until his front gun jammed. Swinging his 'Four' round enabled Venmore to hammer a further 150 rounds into the enemy fighter, which then began to go down trailing smoke which quickly enveloped the machine. However, they had to attend to the other Germans so did not see their victim's demise. Both men were then wounded in further attacks, but Roulstone got them back over the lines and was later able to give a good report of the combat from his hospital bed. All they could claim was an out of control victory, but unknown to them Hans Bethge, commander of Jasta 30 since January 1917, had fallen to his death. He had been flying Pfalz DIII 5888/17. His body was taken home to his native Berlin for burial.

The Germans evened up the score slightly this day by downing Flight Lieutenant R P Minifie of 1 Naval Squadron. Flying Camel B6420 he was in combat with a Pfalz Scout near Houthulst Forest, was hit and had to come down. Richard Pearman Minifie had 21 victories and a vast experience of front-line combat – almost a year had passed since he made his first combat claim – which had brought him awards of the DSC and two Bars. As it happens, his victor was a young Gefreiter by the name of Gebhardt. Although he had been flying with Jasta 47w since December, Minifie was his first confirmed victory. That's how it went sometimes; the not so experienced, or sometimes the vastly inexperienced, just did the business! He would gain a second victory in April, but that was all. Minifie ended his war in a prison camp, having come down south-west of Staden, but at least he had survived, and would live another 51 years, almost to the day.

Captain StC C Taylor of 80 Squadron was another ace lost on 17 March. With ten victories scored flying DH5s with 32 Squadron in 1917, and now a flight commander on this Camel Squadron he had received the MC. This 22-year-old South African was probably shot down by Heinrich Kroll of Jasta 24 west of St Quentin, at 11.00 (G). Taylor, in B9209, was killed. He was last seen at 10.20 diving down on a German machine south of Cambrai and has no known grave. This was Kroll's 21st.

The other McCudden
Captain James McCudden VC DSO MC MM had a brother in France who was also a fighter ace, John Anthony McCudden MC. By comparison his score of eight was smaller than his famous brother, but they had been scored steadily since October 1917 while serving in 25 Squadron on DH4s and later with 84 Squadron, on SE5 scouts. Who knows what he might have achieved if the fates had not been against him on 18 March 1918.

Captain F E Brown MC, who we met previously when Hobson shot down Mettlich, was leading a patrol of 84 Squadron in the late morning, a show along a line St Souplet–Sequehart, escorting DH4 bombers from 5 Naval Squadron, attacking Busigny. No.84 was a superb unit at this time, most pilots having scored something, and most were or would be aces. Today was no exception, for apart from Brown, the patrol consisted of Captain Ken Leask MC, J F Larson, P K Hobson, G O Johnson and W H Brown, as well as Anthony McCudden (B172).

The SE5s made rendezvous with Camels of 54 Squadron at 10.30 over Flez aerodrome and proceeded to Busigny via Hamblières. They reached the target area at 11.00, but they could see no sign of the bombers, so worked their way westwards. Ten minutes later they spotted the 'Fours' approaching from the north-east, being pursued by 30 enemy fighters, Albatros Scouts and Triplanes. The Camels turned east, followed by 84, and the engagement began east of Busigny at 16,000 feet. There was a general mix-up of Camels, SEs, Albatroses and Tripes, the fight lasting around ten minutes, generally working east. A good number of the German pilots spun down; indeed, 84 claimed five out of control with another crashed, while 54 claimed one in a spin and another down in flames. The opposing Germans were von Richthofen's Circus, one victory being scored by Jasta 11's Hans Wolff, his first of ten kills, and this was over the younger McCudden. He fell at Escaucourt at 11.15 and was buried at St Souplet. No.54 Squadron also lost three pilots, two to Jasta 11, one by Manfred von Richthofen himself (victory No.66), and one to Jasta 10, who in turn lost one pilot.

It was not the first time Anthony (Jack) McCudden had been brought down, but it was the last. On 28 February 1918 he had become the victim of Ulrich Neckel of Jasta 12, while flying SE5a C5310. It had been Neckel's sixth victory, of 31, but luckily McCudden had got over the lines before hitting the ground by the Bois de St Gobain.

These actions ended the pre-Offensive period. The German's mighty March attack, which was later called the Kaiserschlacht (Kaiser's Battle), was due to start on 21 March.

CHAPTER EIGHT

March – April 1918

The German's March 1918 Offensive
The German Offensive began at 09.40 hours on 21 March 1918, following a five-hour bombardment of the British lines. Then three German armies attacked the British Third and Fifth Armies' trenches. Like all previous offensives on both sides, it was hoped that this would finally break the awful deadlock of the Western Front. It so nearly succeeded, but finally failed just as miserably as all the others.

Naturally the airmen of both sides became heavily involved and the aces were much in evidence. A problem, however, for the British fighter pilots was that in order to help stem the waves of attacking German troops, many were used in ground attack missions, either carrying small 25lb Cooper bombs or just ground strafing with their Vickers and/or Lewis guns. Thus the men who had gained some little mastery of air combat in the crisp air above the trenches became vulnerable to the curse of ground fire, not only from dedicated anti-aircraft gunners, but to any 'soldat' with a rifle or machine-gun.

Hanstein finds the sting in a Bristol's tail
At first poor weather hampered any attempts by the RFC to fly sorties against the advancing German troops and it was not until the afternoon that any appreciable missions were flown. For once the advance had completely overwhelmed the front-line trenches, making huge inroads into the British Front. While two-seater Corps machines made valiant efforts at directing artillery fire on a very fluid front, Camels, SE5s, the new Dolphins, and Bristol Fighters took to the sky in order to strafe the aggressors.

The Brisfits of 11 Squadron mounted a DOP – Distant Offensive Patrol – away from the front, in order to combat any aircraft climbing towards or descending from the front, or to attack any transport seen on the ground making for the battle zone. It was 16.15 and they were at 17,000 feet making their way back to the west. Then some enemy machines were seen below, near the lines. Lieutenants H W Sellars and his observer, C C Robson (C4673), attacked an Albatros two-seater, whose crew were directing artillery fire. Herbert Sellars, a Cheshire man, and Charles

Robson had been a crew for some time and had already achieved four victories during this month. Sellars fired a burst from above, then as he dropped below the C-type, fired twice more as he zoomed up from below, the last from only 50 yards. The two-seater dived vertically and then went into a spin, Robson seeing smoke coming from the German's engine. Another crew watched it crash to the ground on the British side of the line, north of the Bapaume–Cambrai Road, near Morchies. Ordinarily the wreck might have been inspected by the RFC but the battle moved too swiftly and the area was soon over run.

Sellars meanwhile had spotted another two-seater with a double-tail – most likely a Hannover – attacked and fired five bursts from 100 yards, but the German's only reaction was to head east over the lines. Sellars did not follow as Robson had warned him of approaching fighters, three Albatros Scouts coming down behind them. Sellars put the Bristol into a sharp turn to the left as Robson began hammering away with his rear Lewis gun, two bursts going into the nearest V-strutter. As he watched, the Albatros burst into flames, nosed over and crashed into the ground north of Morchies.

The pilot of this Albatros has always been noted as Bavarian Jasta 35's commander, Leutnant Ludwig Hanstein, who had gained his 16th victory this same afternoon. He was shot down in flames by a two-seater between Noreuil and Vaulx Vraucourt, flying Albatros DVa 5285/17, although, due to the raging battle, his body was not discovered until 2 June. Three German pilots were lost this day, the other two much further south, at Bellicourt and Nauroy. While Hanstein fell in the right area for Sellars and Robson, the German's record his fall at 18.00 hours, and his Camel victory at 17.45, whereas 11 Squadron say their combat was at 16.15. With British and German times the same, can one discover an error in the German timing? The only Camel loss (Hanstein was seen to shoot down one, which was confirmed by his pilots, although a second one was not confirmed, timed at 17.45 and 17.50 near Bapaume and at Bertincourt) that appears to fit Hanstein's 16th victory was a 3 Squadron machine (B9155) that was shot down at Vaulx Vraucourt at 16.00. Leutnant Baron Wolf von Manteuffel, also of Jasta 35b, was also credited with a Camel at 16.00 hours, much more in keeping with the main combat period.

It would seem that Hanstein fell at 16.15, not 18.00, and that his final victory may have been some form of final consolation, which his pilots claimed on his behalf. Obviously, with the combats taking place over the battle area, confirmation would prove difficult for the German pilots. Three other 3 Squadron Camels were noted as casualties this afternoon, with one pilot returning wounded, and two others shot up, but whether by air combat or ground fire is unclear. Sellars and Robson were credited with two kills this day and would add two more to their tally before being brought down on 15 May – see later.

On Day Two of the offensive, three aces fell, an Australian and two Germans. One of the Germans was Leutnant Hans Gottfried von Haebler of Jasta 36, who had eight victories. The RFC and RNAS were not the only fliers engaged in ground attack work during the March offensive, the Germans did so too, and it was von Haebler's undoing. He was brought down by ground fire near Bapaume and was mortally wounded, crash-landing in the British lines. He died of his wounds in a British hospital the next day. His Triplane, 509/17, received the RFC designation G.158 of captured aircraft.

The second German was a Marine pilot, Leutnant zur See Bertram Heinrich of MFJI, who had achieved ten victories to date. He met up with another ace, Flight Commander A M Shook, and came off second best. Shook was leading a 4 Naval Squadron escort mission to a French photographic machine to Ostend in the afternoon and was in the Slype area at 15,000 feet. They were attacked by 15 Albatros Scouts and what was described as a new type of scout with a rotary engine – possibly one of the first Siemens-Schuckert DIIIs to arrive at the front? The Camels turned to engage the hostiles, Shook (B6300) opening fire on one which fell out of control over Slype, which was seen to crash by the French observer. In the subsequent fight, two more German machines were seen to go down, but owing to the intensity of the combat their end could not be watched. One of these was hit by FSL S E Wise (B6243) who noted the machine was painted yellow with black stripes near the wing tips. Later the French confirmed that a second Albatros was seen to fall in flames. Whatever the outcome, either Shook or Wise had hit and wounded Heinrich, who still managed to get his machine down on Vlisseghem airfield. Leutnant Brockhoff was also wounded, but survived to command MFJIII in June. Heinrich too would return to duty, but would not survive the war.

The Australian was Captain Richard Watson Howard, a 21-year-old from Sydney although he had been raised in Newcastle, New South Wales. Having studied engineering he joined the Australian Army Engineers in 1915, served in France, but then transferred to the Australian Flying Corps. He had mostly flown bombers in 1917, with 57 Squadron RFC, but had then become a fighter pilot with 68 AFC Squadron, which became No.2 AFC Squadron, equipped with DH5s but then SE5a machines. On 18 March 1918 he downed his eighth victory, by which time he had become a flight commander and received the MC.

Flying SE5 D212 on the 22nd of March, he took off with Lieutenant A

R Rackett at 16.30 and ran into trouble south-west of Cambrai. Howard had dived on a two-seater below some cloud which must have either been heavily, but cunningly, escorted, was a straggler, or perhaps a decoy. As the two SE5 pilots closed in, they found themselves in the midst of about 30 single and two-seaters. In the ensuing fight, Rackett claimed a two-seater before breaking off and heading for home, but because of ground fog he put down at Candas, staying the night with 49 Squadron. Therefore, he did not know his flight commander had not returned until he flew back to base the next day. He reported seeing Howard south of Honnecourt, while ground observers had seen him above Epehy, but both are close to each other. In some accounts the German force was led by Jasta 79, commanded by Leutnant Hans Böhning. This was a fairly new Jasta, having come to the Front at the beginning of February, Böhning having previously been with Jastas 36 and 76, having scored five victories.

The problem, however, is that Böhning and one of his pilots, Leutnant Buchstett, both claimed a Camel each over Vermand, which is some 15 to 16 kilometres south of Epehy–Honnecourt. In fact, Camels of 3 Naval Squadron were operating near St Quentin this afternoon and two collided in combat. Were these the two Camel claims? But then, Jasta 68 claimed two Camels in the same area, so they might be their victims. It would seem obvious that von Böhning would know the difference between a Camel and an SE5 – not that this is conclusive; Germans often confused the two – but he had previously shot down at least two Camels.

Therefore we have a difference in location and aircraft type. So who might have been in action with the two Australians? The reason von Böhning is credited is that in the Nachrichtenblatt his victory is shown as an SE, as is Buchstett's, although the Jasta records them as Camels. However, that doesn't explain why there are two claims for one loss, nor the assumed difference in location. There are no other obvious claimants, so unless something else comes to light, the nearest one can assume is that the fight drifted south, that Böhning's victory was indeed an SE5, and that Buchstett thought he'd got Rackett.

One other SE5 ace was wounded this day, Captain Paul Beanlands MC of 24 Squadron (D272), on an afternoon sortie. However, he managed to get home, but his war flying was over. He died in a flying accident in England on 8 May.

Beware the balloons
Leutnant Erich Thomas had what might be termed 'balloon fever'. It affected, if that's the right description, a number of aviators in World War One. They seemed obsessed by these dangerous targets. Dangerous because it needed a flight into enemy airspace to get at them and both

sides defended their valuable but vulnerable observation balloons with light and heavy AA guns as well as patrols of fighter planes.

They were, however, important targets. Both sides used them extensively to watch the other side's movements and also to direct artillery fire on men and equipment. Thomas had flown with Jasta 9 and during the first weeks of 1918 had shot down eight balloons. Then he had transferred to Jasta 22, and on 21 March had burned his ninth gasbag, followed by his tenth victory the next day – and his first and only aeroplane. On the 23rd he was back after balloons but this time his luck ran out. He was caught over the front by three French pilots of Escadrille Spa100 and Spa57. Sous Lieutenant Claude Haegelen of Spa100 had achieved two victories to date, although he would end the war with 22. His two companions on the 22nd were from Spa57, Lieutenant Auguste Lahoulle, who had yet to open his account, and Lieutenant Jean Chaput, a long-serving ace with 12 victories. Lahoulle would achieve ten kills before being wounded in July, Chaput 16 before being killed in May. They shot down Thomas over Courcy-le-Château, near Bohain, and he was taken prisoner. In fact, Lahoulle and Chaput then took out Thomas's companion, Leutnant Hans Unger, who was likewise taken into captivity.

Maurice Mealing's mysterious loss
Lieutenant M E Mealing MC was four months short of his 25th birthday in March 1918, and had been with 56 Squadron since the previous year. He had also served in 15 Squadron as an observer in BEs prior to taking pilot training, so he was an experienced airman. With 56 Squadron he had scored 14 victories.

On 24 March he had been part of a nine-man patrol that had flown out to strafe advancing German troops this Sunday afternoon. Not surprisingly the SEs became split up over the front and Mealing and Lieutenant Walkerdine became separated as they chased after a couple of two-seaters they spotted. Walkerdine then experienced gun stoppages and broke away but Mealing carried on, Walkerdine last seeing him over Le Transloy. Although he failed to get home, everyone thought Mealing was safe, even if a prisoner, for someone had seen his SE5 in a field and a pilot who looked very much like Mealing had waved. However, his death was later confirmed, although his grave was never found. It is possible he had attacked the two-seater crewed by Unteroffizier Zetter and Leutnant Tegedor of FA245 and was either damaged by return fire or, being low down, had been hit by fire from the ground. (FA245 in fact claimed a Camel brought down.) If he had been the pilot seen waving, why had he not been taken prisoner? In the heat of battle, and with the almost hourly attacks by low flying British aircraft, did some disgruntled

German soldier open fire on him, or did Mealing look as if he was going to resist? Perhaps he was already mortally wounded. We will probably never know.

Beware the 'Harry Tate'
The British Corps machine, the RE8, also known to its British crews as a 'Harry Tate' after the musical entertainer, was often seen by German fighter pilots as easy meat. It did have a few advantages over its predecessor, the range of BE2s, and in capable and alert hands, it often surprised an opposing fighter pilot who thought he saw a chance of quick glory.

Leutnant Hans Müller of Jasta 18 was one. He had just been involved in the mass change-over of pilots from Jasta 15 to 18, Rudolf Berthold having been permitted to swap them in order for him to have men of his old unit under the aegis of JGII. Müller, a Frieslander, had progressed from two-seaters to fighters as had many other successful fighter pilots, and had achieved two confirmed and one unconfirmed victories. The RE8 crew he spotted over the front on 27 March very nearly ended his scoring run on this day, but he was lucky and he was allowed to go on to score 12 kills, including four American Spads in one day in September. The RE8 crews were always out ranging artillery or taking photos, or flying contact patrols, even in poor weather. No.59 Squadron was one of the many units occupied in this fashion on this day. Its crews were involved in several air actions and it is difficult to be firm on just which crew met Müller. One difficulty is that there is no apparent record of the time of his combat, which would help to establish things.

Lieutenants William Laugham Christian and John Edward Hanning, the latter a Canadian who wore glasses (!), took off at 06.15 (in 7722) to fly a contact patrol and had a combat with a fighter over Grandcourt and claimed it shot down. It seems they were attracted to a nearby balloon and were about to attack it when they were intercepted. According to Hanning's diary, the battle which developed was a fierce one. However, their own aircraft was damaged and they had to force-land at Marieux (Hanning records the location as Couin) and crashed in doing so. They survived allright, but for the 19-year-old Christian, who had been born in India, it must have been a sobering experience, for he had only arrived on the Squadron on the 25th. He was not allowed any practice flights following his arrival and the next day flew his first war sortie, a bombing raid which was unsuccessful due to a forced landing with engine trouble ten minutes after take off.

John Hanning was far more experienced. He was 28 years old for a start, and had been on the Squadron since the previous September.

Although nothing is recorded, he was wounded in the leg during this encounter, two splinters from one of Müller's bullets remaining in his leg for the rest of his life. The actual bullet had exploded inside a map case. Hanning was knocked off his seat but got up and gave the German another burst. His wound was dressed in a field hospital but he refused to stay hospitalised and was back with 59 on the 29th. His tour of duty ended in April and he returned to England. He later became a millionaire. Christian returned to operational flying on 1 April and continued to fly with the Squadron until the end of the war, his last flight recorded on 22 December.

As Müller had claimed the RE8 also shot down, and obviously front-line troops had seen the RE8 crash-land inside British lines, the weight of evidence points to Christian and Hanning being the opposing pair. But Lieutenants E G Leake and T H Upfill had also been involved in a fight in the early afternoon and brought down a fighter into the front line trench area. However, Gilbert Leaske and Thomas Uphill, both on their second periods with 59, were not shot down.

Richthofen's 72nd victory
One problem of being, or some might say trying to be, a historian, is that deductions sometimes prove false, by which time the findings have already appeared in print and therefore seem 'cast in stone'. Without doubt, some of the conclusions in this book will prove in error. It is one thing to make an error when the facts are known, but quite another if the best evidence points to something being so but subsequent new information casts a different light on to it. When Floyd Gibbons was writing his famous *The Red Knight of Germany* in the 1920s, he clouded the waters sometimes by trying to make facts fit the circumstances as he believed he saw them. Sometimes this can be done unwittingly, which is an honest mistake, but sometimes, when it is not, then that is cheating history. I hope in this book that any mistakes are honest ones and not taken as dishonest!

In trying to determine some of Manfred von Richthofen's 80 victory claims, Gibbons had good material to go on, especially when a victim fell on the German side of the lines. The difficulty always was when the opponent came down on the British side, or, as during March 1918, the fighting took place over a major battle area. Gibbons then had to make deductions and hope they were right. On one or two occasions he bent the rules a bit to suit himself. He did not, of course, have access to the information available today.

The problem he faced with Victory No.72 was that he had to rely on von Richthofen's identification of his opponent. What's wrong with that,

I can hear some cry? Nothing I suppose, so long as you believe every pilot in every war is able positively to identify other aircraft in the heat of combat. The fact is that both sides constantly misidentified other aircraft. In combat the Germans often mixed up Bristol Fighters with AWFK8s and DH4s. All pusher types in the earlier years were Vickers, be they Vickers FB5s, FE2, FE8s or DH2s. They would also universally call them all 'gitterumpfs', i.e. lattice-hulls. On the Allied side too, Albatros and Pfalz Scouts could be confused, Fokker DVIIs and Pfalz DXIIs too. The former two would also be referred to simply as Nieuport types, because of their sesquiplane configuration, or V-strutters due to both types having an outer wing 'V' strut.

On this 27th of March, von Richthofen identified his target as a Bristol Fighter, so Floyd Gibbons had to look for a corresponding BF2b which had come down in the battle area and from which the crew had not got back. He had one advantage, however, in that he managed to contact one of the men in the BF2b he chose – Captain J H Hedley. He confirmed he and his pilot, Captain K R Kirkham MC, had been brought down by a fighter, but said he had never heard that Richthofen had been the victor. Whether Gibbons asked Hedley the time of their shoot-down will never be known, but if he had and Hedley had recalled it had been around 10.00, Gibbons should have known they could not possibly have been engaged by the Red Baron, for his claim had been timed at 16.30 in the afternoon.

Captain Robert Kirby Kirkham MC had scored eight victories along with his various observers while serving with 20 Squadron since September 1917, so he was no novice for any German pilot to tangle with. We now know he and Hedley were engaged over the front by Leutnant Karl Gallwitz of Jasta Boelcke, the same pilot who had killed Rhys Davids of 56 Squadron the previous October. His claim is bang-on for time, 11.00 hours, and place, south Albert. Gallwitz would claim one more kill, his tenth, before being injured in a crash in April. So what did von Richthofen shoot down? In all probability it was a 'Big Ack' of No.2 Squadron.

Trollope of 43 Squadron
Captain John Lightfoot Trollope MC and Bar ran up a remarkable string of 18 victories in the first three months of 1918, including no less than six on 24 March. On the 28th, 43 Squadron provided ten Camels for a protection mission for ground strafers, taking off at 08.10. One pilot had to abort but he had his aircraft fixed and caught the others up near the front. The Flights were:

Capt J L Trollope	C8270	Lt R J Owen	C8259
2/Lt W J Prior	D6404	2/Lt C R Maasdorp	C8224
2/Lt H T Adams	C8267		
Capt L G Loudon	C8215	2/Lt C F King	D1777
2/Lt A C Dran	D6428	2/Lt M F Peiler	D6452*
2/Lt O'Neill			

When the Camels began to drift back, the ground crews noted with some concern that five were missing, including the indomitable John Trollope and Robert Owen, another of the Squadron aces.

There were two main areas of fighting on this morning, 43 Squadron in a scrap east of Albert while 2 and 4 AFC Squadrons, along with 46 Squadron, were heavily engaged further north, east of Arras, with Jastas 14, 52 and 32b. The Camels of 43 were intercepted and engaged by aircraft of Jasta 4, 5, 11 and 56, and the scrap seems to have split into two groups, one moving north, the other south. The fight started soon after 09.00 and lasted for half an hour.

Fritz Rumey of Jasta 5, Rudolph Heins of Jasta 56 and Victor von Rautter of Jasta 4 all brought down Camels along the Somme, south-east of Albert (Bray sur Somme (09.10), Morcourt (09.25) and north-east of Suzanne (09.30)), while to the north Ernst Udet of Jasta 4 shot down Maasdorp east of Thiepval at 09.10. Kaleta of Jasta 56 shot down his near Chaulnes, much further south, at 09.20. In any event, Adams and Maasdorp were dead, the latter having gone down under Udet. Trollope, Prior and Owen were all prisoners. Cecil King saw Trollope fighting eight scouts and went to his aid, shooting down one of two he managed to draw off, but he was then shot up and wounded and had to break away for home. In addition, Loudon and O'Neill had their Camels badly shot about.

It has been recorded in the past that Paul Billik of Jasta 2 claimed Trollope, but he in fact shot down 'his' Camel at 10.00 at Sailly-en-Ostrevent, east of Arras, piloted by 2/Lt C Marsden of 46 Squadron (C1649). Two 2 AFC SE5s that were also lost in the Arras fight were brought down by Jasta 14 and Jasta 32b, at Gavrelle and Mercatel. Who exactly brought down Trollope and Owen is unclear, but it was either Rumey (his 13th of 45), Heins (his third of four) or von Rautter (his first of 15); but probably Rumey.

Other than King's claim, there were no others until Trollope returned from Germany, repatriated in October, having had to have his left hand

---

* Aborted and took off again.

amputated. (He later had to lose the whole arm up to the shoulder.) In his report he claimed to have destroyed a balloon, then shot down two Albatros Scouts before falling himself. He also claimed a victory for Owen, which was credited as his seventh. In the event, only one Jasta 4 pilot was wounded in this fight, Leutnant Georg van der Osten, in Pfalz DIII, 4566/17, an ace with five victories when he'd flown with Jasta 11 in 1917. He did not see further combat.

April 1918 – the Offensive peters out
The German Offensive, after doing so well, ended on 5 April. Ground attacks had and did continue, and indeed were now a well established tactic by the British fighter force in France, and would continue to be so until the Armistice. Camels particularly became adept at ground strafing.

The day before the end, 4 April, Leutnant Heinrich Geigl the Staffelführer of Jasta 16b, was killed. In one report he collided with a Camel of 65 Squadron flown by Second Lieutenant D Kennedy (D6552) during an air fight over Warfusée at 15.30, Geigl's comrades crediting their lost leader with his 13th victory. Oddly enough, there is another good claimant, Captain F M Kitto of 54 Squadron. Kitto, in Camel C1603, was leading an Offensive Patrol south-east of Hamel (Warfusée is south-east of Hamel) and at 15.30 he observed about eight enemy fighters above and engaged one. After sizing each other up for a short while the German pilot suddenly dived away east and Kitto chased after him. Firing a long burst from 100 down to just 30 yards, the machine continued down in a fairly steep dive and flew into the ground, crashing badly. Second Lieutenant C S Bowen confirmed seeing the wreckage in flames on the ground.

On the face of it it really depends on whether the story of the collision is true, but it certainly seems so. So perhaps the German pilot who crashed after Kitto's attention was someone else who wasn't even injured; there was only one other German pilot killed this day, a Jasta 46 pilot who crashed near the airfield at Liéramont, 34 kilometres to the east-north-east of Warfusée and Hamel.

Lieutenant W H Sneath, from north London, had just reached his 19th birthday. Flying with 208 Squadron, which until a week earlier had been 8 Naval, he had scored four victories while still only 18. On 6 April the Squadron had flown out on an OP to Lens at 10.00, Wilfred Sneath in B7187. The eight Camels fought a battle with several enemy fighters, the other pilots seeing the youngster send down a Fokker Triplane, out of control, but he failed to emerge from the fight. He was brought down by Leutnant Karl Hertz of Jasta 59, his second of three victories he would score before his own death on 9 May.

Another ace for Richthofen

Manfred von Richthofen brought his score to 76 this afternoon of 6 April. The British fighters were still hounding German troops with bombs and machine-guns in the battle areas, and if caught low down, were at a disadvantage. Von Richthofen caught some of 46 Squadron's Camels at low level north-east of Villers Bretonneux, and shot down Captain S P Smith (C3919). Smith had five victories. He had flown two-seaters and had been a flight commander with 46 for exactly one month. He was known for his aggressiveness, but this day the Baron got him before he had a chance at real combat. Although his father later found the area where his son had crashed he was unable to locate any sign of his rough grave.

On 7 April, Captain G B Moore, a Canadian with 1 Squadron who had scored ten victories since the previous October, was killed. Often noted as one of von Richthofen's victims, he was in fact killed by a direct hit from an AA shell. He was not in the area of the Baron's combat, nor does the time tally. Guy Moore's SE5 (C1083) fell at Hollebeke, up on the northern sectors, not down south on the Amiens front.

At least Captain W D Patrick survived his shoot-down on 10 April, to become a wartime guest of the Germans. He had accounted for seven German aircraft while flying with 1 Squadron from the late summer of 1917, flying both Nieuports and SE5s. The Ayrshire man was a lot older than Sneath had been, 28, but on a low-flying patrol to the Armentières area, leaving at 09.40 (SE5a B8371), he didn't get back. Although he was seen in combat with an enemy machine there is no obvious victor, and two of his companions were brought down too, both being wounded by ground fire. One was an American, Francis Peabody Magoun. He had four victories thus far but the New Yorker returned to the Squadron later in the year and bagged his fifth victory in October.

Walter Göttsch is killed

Yet another German pilot ran foul of an RE8 on this same 10 April. For Walter Göttsch it was virtually an exact re-enactment of the fight involving Hans Müller on 27 March, except that Müller survived.

Lieutenants H L Taylor and W I E Lane were operating near Thennes at 12.45, flying at 1,500 feet, on what was loosely termed a situation patrol. Nine Fokker Triplanes of Jasta 19 came tumbling out of low cloud, the leader opening fire at about 100 yards. Lane grabbed his Lewis gun and began to return fire and was dead on target. He saw the Triplane stagger, then plummet down in a spin over the front lines. Another

Triplane also fired and Lane caught a bullet in the leg, but continued to fire back, claiming a second Triplane hit, but this one burst into flames and went down over the Bois de Gentelles. Front-line German troops saw the Triplane crash but also saw the RE8 start to go down, for Taylor was forced to make a rapid landing due to battle damage. This is why Göttsch was credited with his 20th victory – RE8 B6641 – although it landed on the British side of the trenches. The German Staffelführer too came down by the British lines, hence the wreckage of 419/17 being given the serial number G.163.

Old hand – new game!
The next day, 11 April, Captain Kelvin Crawford, an old hand, was lost. He had seen action back in 1916 on DH2 pusher scouts with Lanoe Hawker's 24 Squadron. By the time he went home for a rest in July 1917, this south Londoner had achieved five victories. By the time he returned to France in March 1918, posted to 60 Squadron, the war in the air had changed dramatically, and in any event returning to the front was always a dangerous time, however experienced a pilot was. One developed a sort of inbred skill and, having survived for a period, a pilot would be, as the younger generation might term it today, 'with it' or perhaps more accurately, 'street wise'. It took a little time to hone that skill once more, and until a pilot did so (got back into the swing), he was tremendously vulnerable.

Kelvin Crawford took off on patrol at 15.35 in SE5 C5445 and failed to return. As far as one can see, 60 Squadron met Jasta 5 to the west of Bapaume, and at 17.15, Leutnant Otto Könnecke shot him down over Bucquoy. It was the German's 17th victory and he would go on to score 35 by the war's end.

How did Albert Dietlen meet his end?
It is not easy to tie up loose ends in the who-got-who jigsaw and a case in point concerns the loss of Albert Dietlen, the Staffelführer of Jasta 58. Bavarian-born Dietlen had been actively flying since 1916, first as an observer, then a pilot. As an air fighter with Jastas 23b and 41 he had raised his score to seven in March 1918, then was given command of Jasta 58, on 4 April. Although it had been operational since January 1918, it had yet to achieve its first victory.

Dietlen himself and Leutnant Spille changed that on 11 April, with two Camel victories, but the next day proved the last for the Bavarian leader. The confusion over his loss is caused by some records saying he collided with an RE8, while the Nachrichtenblatt notes a victory awarded over a Bristol Fighter. So again we have the dispute between what type of machine was in the combat. Added to this there does not appear to be any note of an RE8 being in collision with a German fighter, nor are there any Bristols lost in mid-morning, the time of the action being 10.35. Dietlen is supposed to have gone down over Le Petit Mortier and the main fight in that area at that time involved the Camels of 43 Squadron. They were in action with Albatros and Pfalz Scouts at 10.30 over La Gorgue, a former base of theirs before the recent German Offensive. Considering the number of aircraft 43 is supposed to have knocked down in this fight, it is a wonder the German fighter force in this sector survived at all!

Claims for victories went to Captain H W Woollett, three (two in flames, one crashed; he would claim a further three destroyed in the late afternoon); Lieutenants G G Bailey, one destroyed; C F King, one destroyed; C C Banks, one destroyed; and H C Daniel, two destroyed. This makes a total of eight, not bad considering there were only eight in the first place, and only Dietlen was lost. Hector Daniel, after sending one down on fire, attacked another, which he reported had just shot down a Camel, and sent this down in a spin, seen to crash by Woollett. Admittedly it is difficult to know if the Germans lost aircraft without any injuries to pilots, but not too often did pilots crash without some form of recorded injury, especially the ones in flames. In the event, 43 Squadron did lose two aircraft in this action, one pilot being killed, another taken prisoner. One was claimed by Vizefeldwebel Friedrich Ehrmann of Jasta 47w, who obviously joined in the scrap, another by Leutnant Hävernik of the same staffel, which came down at Le Petit Mortier. Jasta 47 record no pilot losses, although Ehrmann had his machine damaged, forcing him to crash-land his Pfalz DIII near Frelinghem. Perhaps this was Daniel's victim.

So, with no mention of an RE8, or a Bristol, and the two Camel losses, one of which was seen by Daniel, accounted for, Dietlen certainly doesn't seem to have scored his tenth and final victory in this particular air fight which cost him his life.

NB. The times changed again on the Western Front. From 16 April the German time was once again one hour ahead of the Allied time, and remained so until 16 September.

21 April 1918
No self respecting WWI historian or enthusiast will not know why this date is recorded here. It was the day Baron Manfred von Richthofen was shot down atop the Morlancourt Ridge, starting a debate on who got him which has lasted for nearly 80 years thus far.

It may seem churlish to ignore for the most part this particular controversy. Suffice it to say that the weight of the evidence points to him having been hit by ground fire as, with both guns jammed, he must suddenly have realised the perilous position he had uncharacteristically put himself in and pulled up to return to the east. At that moment he was hit by a single bullet from the ground, tried to get down but lost consciousness from his terrible mortal wound as he was about to do so, and breathed his last.

There will no doubt be more written about this episode, not least by this author in a future book under research (and using eye witness accounts which have lain dormant for 60 years). One thing, however, is certain: the previous day, the 20th, von Richthofen had downed two Camels of 3 Squadron to bring his total to a round 80. The first one had been flown by yet another ace, Major R Raymond-Barker MC, the Squadron commander. All his six victories had been scored in 1917 in Bristol Fighters, so he was undoubtedly at a disadvantage when von Richthofen attacked him while flying a Camel. He crashed in flames (D6439).

There was another famous fighter ace in the wars this day too, but he survived to fight another day. Captain Robert A Little DSO and Bar, DSC and Bar gained his 44th victory this afternoon, on a score sheet he had begun in late 1916. Flying with 203 Squadron, he had not had a good day. At 09.30 he had taken off on a Special Mission in B6319 with four of his Flight but had had to make a forced landing north of Lillers; he managed to get his engine going again and completed his patrol.

After lunch he was out again and the patrol ran into Jasta 47w, who had scored against 43 Squadron on 12 April. Little had shot up the Pfalz DIII flown by Unteroffizier Erich Kauffmann, who was so severely wounded that he died the next day. However, in doing so Little left himself open to an attack by Friedrich Ehrmann, whose fire scored hits on the Camel. Little managed to get over the lines but crashed near Lillers again, writing off the machine although he himself emerged unhurt. Ehrmann, however, received confirmation of his fifth victory.

Thus a small fish had got a bigger fish; it was not an uncommon occurrence. It only took a moment's diversion to have someone get you. If Daniel had shot down Ehrmann on the 12th, he had been caught exactly the same way, finishing off a Camel, and allowing another to come in behind him.

NB: It is believed that Captain T S Sharpe DFC, 73 Squadron, was brought down by Manfred von Richthofen on 27 March and taken prisoner. He had six victories.

CHAPTER NINE

April – May 1918

Ken Junor's only day as flight commander
Tuesday, 23 April 1918 started dull, with rain and low cloud, but with a brighter afternoon 56 Squadron took off on patrol at 16.40. Although led by Captain E D G Galley MC, with Junor, W R 'Sambo' Irwin, Cyril Parry, E R MacDonald and John Doyle, Ken Junor had this day been notified that he was the new A Flight Commander.

A Canadian from Toronto, Junor was 23 and had transferred from the Canadian Machine Gun Corps in 1917 following action in France. In the last month of that year he had joined 56 Squadron, gained eight victories and been awarded the MC. Flying east of Fricourt, the SE5 patrol attacked five Albatros Scouts they found over Becourt at 8,000 feet at 17.00 hrs but after an indecisive bout of sparring the SEs headed north. Then Galley dived on a two-seater over Achiet-le-Grand and Junor followed this machine down until both machines became lost in ground mist, and then rejoined the others. Galley then attacked five Albatros Scouts south-east of Bray, along with Junor, Irwin and Doyle, but seeing ten Fokker Triplanes approaching from the south, 2,000 feet above, they all turned north with the exception of Junor, who continued to engage the Albatros Scouts. 56 had been flying with a three-man top cover section above and these engaged the Triplanes without result, but it took the heat off of Galley and his patrol.

Galley spotted either the same Albatros Scouts or others, and attacked a blue-tailed machine over Fricourt which rapidly dived eastwards, Galley following. The Albatros went into a dive and was swallowed up in the mist and low altitude. Galley then pulled up and decided it was time to head for home.

When he landed he discovered Ken Junor was not back, nor did he return. No one had seen him after the Triplanes had arrived on the scene, while still engaging the Albatros Scouts in his SE, C1086. Leutnant Egon Koepsch of Jasta 4 was the only German pilot to claim an SE5 this day, although it is timed at 06.55. However, as there were no actions in the morning, this has to be a recording error for 6.55 p.m. – 18.55. As Junor was lost at around 18.00 British time, south-east of Bray, a victory at 18.55 German time, north of Sailly-Laurette (by the Somme), fits. This was

APRIL – MAY 1918

Koepsch's second victory, his first having been scored back on 5 December. He would not score his third until July but by the end of the war would have achieved nine kills. Oddly enough he was to account for John Doyle too, in September, by then a Captain in 60 Squadron (see later).

The Colville-Jones brothers

A feature of both the twentieth century world wars was the willingness of men to come from all corners of the world to fight for what they thought was right. Robert and Thomas Colville-Jones were two Anglo-Argentinians who lived in Buenos Aires, the sons of a Hurlingham family. Returning to Britain was a natural choice to help defend her way of life. Robert, the eldest, had sailed for the Mother country in September 1914, and served with the King's Royal Rifles Corps and 13th Rifle Brigade while Thomas came in February 1917 and joined the Artists' Rifles prior to being commissioned into the Royal Engineers, having worked on the Central Argentine Railway. Both had later transferred to the RFC, Robert, having won the MC, to become an observer, Thomas a pilot.

Thomas had flown with 20 Squadron on Bristol Fighters from the autumn of 1917 and by February he and his observers had accounted for seven enemy planes. He was then made a flight commander in 48 Squadron, raising his score to 11. Flying with 77937 1st Air Mechanic F Finney, from Hinckley, Leicester, on 25 April they flew on an OP to Villers-Bretonneux (B1126) taking off at 16.05. They were last seen in combat with enemy aircraft and failed to get back. Both were taken prisoner, but Colville-Jones had been badly wounded, injuries from which he died on 24 May.

Only one Bristol Fighter was claimed this day, by a Jasta 5 pilot, Vizefeldwebel Kurt Kressner, although no time is recorded. However, his victim came down near Marcelcave, just to the south-east of Villers-Bretonneux where the fight took place. It was the German's first and only victory, but an important one over such a pilot. Kressner was seriously injured on a practice flight just three days later, in which he broke a leg.

And Robert Colville-Jones? He flew with 57 Squadron, which had DH4 bombers. Flying a mission in A7652, on 4 November 1918, he and his pilot were brought down over Mormal Wood by another Jasta 5 pilot, Otto Könnecke, the German's 34th victory. Another week and he would have survived to return to Argentina.

Heinrich Bongartz's war ends

If Tom Colville-Jones was brought down by a novice, so was the Pour le Mérite winner and commander of Jasta 36 on 29 April. Leutnant Heinrich

Bongartz was a 26-year-old Westphalian who had received most of the honours Germany could bestow upon him. A former school teacher he had been in the infantry at the beginning of the war, seeing action at Verdun as a Sturmoffizier, commissioned in March 1916. Becoming a pilot in the Air Service he had flown both two-seaters and single-seaters and in the spring of 1917 had joined Jasta 36. After scoring 11 victories he had been wounded on 13 July, the first of five wounds he would receive in air actions. Despite these injuries, which occurred in November 1917, 30 March (by German flak), 25 April (slight graze by a bullet), and finally on 29 April 1918, he achieved 33 victories.

On the 29th he was operating near Ploegsteert in Fokker Dr1 575/17 and spotted a patrol of SE5s flying south of Dickebusch Lake. The SEs were from 74 Squadron, led by one of the best patrol leaders of the war, Captain Edward 'Mick' Mannock MC and Bar, who had already achieved 19 victories by this date. One of his patrol was Captain C B Glynn, formally of the Liverpool Regiment where he had held this rank. He had yet to make a claim for a German aeroplane.

Glynn was flying on Mannock's left and saw the enemy machines approaching from his left rear while trying to gain height over the British patrol. Glynn zoomed upwards in a climbing turn towards the leading Triplane, which he noted as green and silver, and both machines went for each other head-on. Glynn began firing with a long burst from his Vickers gun and a short burst from his top-wing Lewis. As the two machines shot by each other, Glynn turned to see it going down, later confirmed by Mannock and Lieutenant H E Dolan MC as having crashed just to the south of the combat area, after diving vertically. Mannock also claimed a fighter shot down as did Dolan. In the head-on encounter, Bongartz had received a terrible wound. A bullet had hit him in the left temple, then passed through the eye and nose. He crash-landed at Kemmel Hill at 11.40 (British time). Despite losing his left eye, Bongartz remained in service, although not as a combat flier, although he was wounded again, fighting against the communists in post-war Germany. It was this wound that finally ended his military career. As for Clive Glynn, he went on to achieve eight victories by the war's end, received the DFC and became a flight commander.

May 1918

Another German ace to fall as May began was Leutnant Hans Weiss, of Jasta 11. Weiss had just passed his 26th birthday (19 April) and flew a Triplane with white rear fuselage and upper wings and tail surfaces as might befit his name (545/17). On 2 May Jasta 11 ran once more into 209 Squadron's Camels, who had been in on the fight which ended in Manfred von Richthofen's death in April.

APRIL – MAY 1918

Lieutenant M S Taylor was in one of the Camels (D3326) which dived on eight Dr1s and fired about 20 rounds into one which went down on its back out of control and was seen to crash by other pilots of 209 and also of 65 Squadron. Weiss had 16 victories when he died, having flown with Jastas 41, 10 and 11. Merril Taylor, a Canadian from Saskatchewan, was almost exactly a year younger than Weiss, who had become his fifth victory. We will meet him again in a later chapter.

As a matter of interest, Jasta 11 lost a second ace this day: Leutnant Edgar Scholtz, in Fokker Dr1 591/17, who crashed while taking off from their base at Cappy at around the same time that Weiss was shot down. Scholtz had achieved eight victories since August 1917, at which time he was with Kest 10, only joining Jasta 11 in January 1918.

Two Frenchmen down
The first week of May saw Sous Lieutenant Omer Paul de Mouldre of Spa84 killed on the third, and Lieutenant Jean Chaput of Spa57 on the sixth. De Mouldre came from Cambrai and had joined the army in 1912 when he was 20. He transferred to the air service, firstly as a mechanic, then an observer and finally as a pilot. Holder of the Medaille Militaire, then the Légion d'Honneur plus the Croix de Guerre with seven Palmes and two étoiles, he had 13 confirmed victories over the period September 1915 to April 1918. On 3 May he attacked a two-seater near le Monchel but was hit by return fire and crashed. Two German crews that claimed Spads this day were Vizefeldwebel Bischoff/Gefreiter Haase of Schutzstaffel 8, and Leutnant Selig/Unteroffizier Gerstenkorn.

Parisian Jean Chaput was another long-serving officer who had entered military service pre-war and had gone through Saint Cyr in 1914. He had been scoring victories since 1915 and on 21 April 1918 had brought his score to 16. He had won the Medaille Militaire, Croix de Guerre with 11 Palmes, one étoile de vermeil and one d'argent, and had been made an Officer of the Légion d'Honneur. Operating from Tille, north of Beauvais, he was flying over the front of the 18 Armee on the sixth and was in combat with enemy fighters. Wounded in this action, he came down inside German lines, and died in an aid station shortly afterwards. He was brought down by Leutnant Hermann Becker of Jasta 12, north of Montdidier, the German's tenth victory of an eventual 23.

Captain C C Clark of 1 Squadron was wounded and taken prisoner on 8 May following an encounter with Leutnant Harry von Bülow-Bothkamp of Jasta 36. 1 Squadron's SEs were on patrol over Kemmel, having taken off at 06.15 and Cecil Clark, from Norwich, was last seen north-east of this location at 07.45. His SE5 (B8410) was von Bülow's fourth victory and he would gain two more before being taken off operations. Following the death of his older brother Walter his mother had requested the surviving son, now heir of the family estates at Bothkamp, Holstein, not to be exposed to further danger, which eventually the authorities had to accede to.

This same day, Roderick McDonald, a Canadian in 208 Squadron who had gained seven of his eight victories while the unit was still 8 Naval Squadron, was brought down by a pilot in Jasta 43, at 12.30 German time. Offizierstellvertreter Josef Trotsky had seen the British and engaged them. McDonald (D1852) had taken off with others to fly a Special Low Recce sortie to the Meurchin–Provin area at 10.30. The German fighters came down on them at 11.15 and a fight developed. McDonald was last seen being attacked by two German machines, Trotsky bringing his victim down south of Auchy. It was the German's second victory, but his last, as he was killed in action over Carvin on 17 May.

One of Mannock's Flight goes down
Henry Eric Dolan had been something of a protégé of Mick Mannock since 74 Squadron had arrived in France. Always in the thick of the action alongside his flight commander, Dolan, a former artillery officer and MC winner, had claimed seven victories in a month.

On 12 May the Flight was in a battle with Pfalz and Albatros Scouts near Wulverghem in his SE5 (B7733). Lieutenant J I T Jones saw what he thought was Dolan's machine go spiralling down into the village itself. They had taken off at 17.35 and green and yellow coloured Albatros Scouts had been encountered at 18.20. Jones had shot one off Ben Roxburgh-Smith's tail which went down in a spin, then he saw Lieutenant Giles shoot one down which crashed. Jones then saw pieces of one fluttering down after Mannock's attack, one of eight Mannock had spotted flying north from Armentières. However, as Mannock attacked, two had collided and fallen to bits, and he had then engaged a Pfalz DIII which dived vertically before crashing into the ground.

Therefore, with four crashed and one out of control, it seems 74 left the field with the honours, even if it did lose Dolan. There does seem to be a slight problem, however, as there do not appear to be any German fatalities this day. Jasta 20's Leutnant Raven Freiherr von Barnekow claimed an SE5 down south of Dickebusch at 19.00 hours, which fits for place and approximate time, but although the Jasta records indicate the date as 12 May, the Nachrichtenblatt shows this to be the 11th, with no claims at all on the 12th. Nor does Jasta 20 note any losses on the 12th.

Walter Ewers tangles with 65 Squadron
Oberleutnant Walter Ewers was just four days past his 26th birthday on

15 May. A former artillery man he had scored eight victories while flying with Jasta 12 and then as Staffelführer of Jasta 77. With Jasta 12 he had flown on the British Front, while with Jasta 77 he had operated mainly on the French Front, but in April the unit had moved to Foucaucourt and was flying patrols over the southern end of the British Front on the Somme.

This day he led a patrol off, flying Albatros DVa 7220/17, and shortly after 09.30 German time he spotted Camels over Villers–Bretonneux. These machines were a patrol of 65 Squadron's B Flight on an OP led by Captain L W Whitehead. South-east of Villers–Bretonneux, at 14,000 feet, they met the six Albatros Scouts, but no sooner had they engaged than several more German fighters came diving down to join in, making a total of around 20 hostiles. In the fight, Second Lieutenant S W Crane (D1791) fired about 400 rounds in a series of bursts into one closing right in to 30 yards' range. He watched as the Albatros turned on to its side, then fell into a vertical nose-dive which continued for 10,000 feet, but lost sight of it when it went into a wood. Meanwhile, Second Lieutenant G D Tod (D1921) had circled around as the fight started until he spotted a favourable opportunity of attacking an Albatros. He fired a long burst into it from 100 to 50 yards and saw it go down but could not watch it as other enemy fighters were trying to get on to his tail. However, Whitehead said he saw this machine go down and shed its wings, while two other pilots saw an Albatros fall to pieces in the air. Captain Whitehead also claimed an Albatros shot down.

These could, of course, be one and the same aircraft seen by the various pilots; Jasta 63 did lose a pilot in the Amiens area, although there isn't a time to help identify the combat. This was Leutnant Walter Lutjöhann, who was seen shot down by a Camel. In any event, what is known is that Ewers was brought down by either Whitehead, Crane or Tod of 65 Squadron. George Tod was an American flying with the RAF and this was his second victory; he would go on to become an ace by August. For Crane, his victory was his first solo effort, although he had helped bring down a two-seater inside British lines a few days earlier. Lewis Whitehead received credit for his fourth victory this day. He gained one more before his own loss (see later).

Not a good day for the Bristols
This same 15 May saw a real mix-up between Bristol Fighters and German scouts, 11 and 48 Squadrons taking a bit of a beating. In all, seven BF2b machines were claimed by the German pilots, although only four were actually lost, with a fifth down inside Allied lines. Others were shot up, but once again there was probably some misidentification of machine types.

The first Bristol claim was at 12.50, by Jasta 6, but this was probably a 57 Squadron DH4. The real crunch came at 13.00 British time, as 48 Squadron got into a fight over La Motte with JGI – Jastas 6 and 11. The Brisfits had taken off at 13.00 and ran into the Triplanes shortly after two o'clock.

They lost two machines, one of which was piloted by Captain C G D Napier MC/Sgt P Murphy (B1337), Charles Napier being an ace with nine victories, all scored with observers other than Pat Murphy. They came down in flames over La Motte, which is just north of Caix and Guillaucourt, locations where Leutnants Hans Wolff of Jasta 11 and Hans Kirschstein of Jasta 6 downed Bristols for their tenth respective victories. Kirschstein would go on to score 27 kills, but this proved to be Wolff's last. Napier, from west London, and Patrick Murphy have no known graves.

The other 48 Squadron loss was C855, Second Lieutenants C L Glover/J C Fitton, who were also killed. Two other Bristols were shot up by the Triplane assault, returning with wounded on board. Leutnant Erich von Wedel of Jasta 11 also made a claim for a Bristol south-east of Guillaucourt. All three claims were made at 15.10 to 15.15 hours. Lieutenants E R Stock/W D Davidson claimed two Triplanes down in flames, but certainly JGI suffered no pilot losses.

A couple of hours later, 11 Squadron sent out a patrol to the south-east of Albert, taking off at 15.40. They also ran into Fokker Triplanes, Jasta 6 again, and Jasta 5. They picked off two ace crews, Liverpool-born Lieutenant H W Sellars MC (killed), with Lieutenant C C Robson MC, taken prisoner (C845), and Captain J V Aspinall/Lieutenant P V de la Cour, a South African (C4882), both being killed. Hans Kirschstein brought his down at Contalmaison at 18.20(G), the Jasta 5 pilot, Josef Mai, crashed his east of Ovillers five minutes earlier. For Mai, this was his 12th victory. Napier and his observers had scored over nine German machines, Aspinall over six. However, both their final victories were a shared Triplane 'out of control', a claim submitted by the returning crews on their behalf.

Captain J S Chick MC and his observer Lieutenant E C Gilroy (C797) claimed two Fokkers and a Pfalz in this fight (to add to a two-seater they had claimed soon after dawn), but again it is not possible to trace corresponding German loss of pilots. Another crew claimed a Pfalz out of control.

Who brought down Hans Wolff?
The day after shooting down his tenth victim, Hans Wolff of Jasta 11 died in action. He came down south-east of La Motte, on the German side of

APRIL – MAY 1918

the lines, with two bullet wounds below the heart. He was later buried in the Heroes Cemetery at Cappy.

There was a mass of air fights on 16 May, several with Triplanes, but if the time given for Wolff's fall is correct, at 08.20, then it is difficult to find a British claim this early. Over the years, the nearest as far as location is concerned was that of Lieutenant H D Barton of 24 Squadron. He claimed a Triplane downed south of Proyart, which is roughly the area, but at 19.15. If, of course, Wolff's time was actually 20.20 (G), i.e. 8.20 in the evening, then this may solve the time problem. However, a Jasta 4 pilot was shot down at Rosières at 20.00, a location which is eight kilometres south of Proyart, if one stretches the time a bit.

Wolff, who had downed Jack McCudden on 18 March, was 23 years old. If Barton was his victor, this 26-year-old South African ended the war with a DFC and Bar as a reward for his 19 victories.

This same day, 16 May, 56 Squadron lost one of its stalwart fighter pilots, Captain Trevor Durrant. He had started his war flying in the back seat of a DH4 with 55 Squadron, gaining one 'out of control' victory in May 1917. Becoming a pilot he was assigned to 56 Squadron a couple of weeks before Christmas and in the first months of 1918 brought his personal score to 11. It was an odd occurrence that he, like Ken Junor, was made flight commander only on the day of his death. Flying a Combined OP over the 3rd Army Front, eight SE5s took off at 18.50, with Captain Louis Jarvis leading pilots of C Flight, Durrant (B183) the pilots of B Flight. They crossed the lines over Albert, having climbed to 13,000 feet and forty minutes later found and attacked a formation of six Fokker Triplanes over La Sars. Durrant and his men were above C Flight as cover.

Initially, Jarvis and his pilots did not have sufficient height to engage so climbed towards the lines. Cloud then separated the two SE formations, and Durrant, thinking that Jarvis would be back shortly, decided to engage himself, but Jarvis and company failed to find the action. The three B Flight men separated slightly as they picked individual opponents. In the fight which developed, one pilot had to break off with a gun jam and then a con rod broke, but he got home. A second pilot was hit and spun away with a damaged engine but got back over the lines and made a successful forced landing. However, both pilots had seen an SE going down in flames; that could only have been Trevor Durrant. The Triplanes were from Jasta 6, Hans Kirschstein claiming an SE5 down at Contalmaison at 21.10, for his 13th victory.

No.62 Squadron lose two aces

This was the second victory of the day for Hans Kirschstein, for he had downed a 62 Squadron Bristol Fighter (C4859) soon after lunch. The crew was Lieutenants C H Arnison MC and C D Wells MC, on an OP to the Corbie area, taking off at 12.15. They were engaged by Triplanes, Kirschstein coming in close to open fire, hitting Charles Wells, who died at his gun, and wounding Arnison. Charles Arnison managed to get his damaged machine back across the lines and made a forced landing near Sailly-le-Sec. He was helped from the machine which was then destroyed by shell fire, but troops got him to a hospital. He had scored nine victories but he did not see further combat.

Three days after these events, 19 May, an ace observer failed to get home. Hugh Claye had been a Captain with the Notts and Derby Regiment but had been seconded to the RFC in March 1918, serving with 62 Squadron where his usual pilot had been the successful Captain G F Hughes MC. As we read earlier they had been involved in the action in which Lothar von Richthofen had had a lucky escape on 13 March. Hughes had now ended his period of war flying and Claye was flying with Lieutenant H A Clarke this day, in C4630. They were part of an OP operating near Corbie, having left base at 08.15 and flying above the formation. Shortly after 10.00 they became embroiled in a fight and were last seen going down under control, near Corbie, at 10.15. They and another two crews were reported missing, with a fourth lucky to survive.

The cause of their losses had been the Bavarian Jasta 34b, commanded by Oberleutnant Robert von Greim, Jasta 5, and flak fire. The Bristols entered the grinder, pounced on by Jastas 34, 68 and 5. Von Greim shot down one which landed near Guillaucourt at 10.10, believed to be C796, with a wounded pilot and dead gunner, his 14th victory. Vizefeldwebel Max Kahlow brought down C4751, putting its crew into captivity, the pilot also wounded, at 10.15. This made it five kills for this 24-year-old. Leutnant August Delling would become an ace, but on this day he had just one confirmed victory. He attacked a Bristol, but so too did Leutnant Fritz Pütter, Staffelführer of Jasta 68. In the event, neither pilot was credited with its demise, as it was also claimed by a flak unit and the disputed victory was eventually awarded to the AA unit. This seems to have been Clarke and Claye, although Kahlow may have been their victor.

Finally, as the fighting drifted west as the remaining Brisfits fought their way back, Jasta 5's Otto Könnecke knocked down B1336, flown by Lieutenants D A Savage and E W Collins. They came down near the front and were unhurt, living to fight another day. Könnecke originally claimed a DH9 but again there appears to be a difference of identification. Fritz Pütter also shot down an SE5 of 84 Squadron whose pilot tried to help out, this time the German's claim being upheld.

Douglas Savage claimed an Albatros DV in this fight, as out of control. It was his sixth victory and was probably Leutnant Karl Bauernfeind of Jasta 34b, who was wounded in the battle. Savage received the MC, claimed one more victory and was then rested.

Richard Plange and the 'Big Ack'
This morning was proving a costly one for both sides. Shortly before the above action between 62 Squadron and the three Jastas, seven-victory ace Leutnant Richard Plange had been killed. The 25-year-old had joined Jasta Boelcke back in September 1917 and by April 1918 had achieved his seven kills and been made Staffelführer of Jasta 36 on 16 May. He had lasted just three days.

The Armstrong Whitworth FK8 two-seater Corps aircraft of 10 Squadron, which he spotted over the front, was from No.10 Squadron (B3315), crewed by Lieutenant W Hughes and Second Lieutenant F C Peacock MC. They were operating on the northern part of the front, up around Zillebeke, having taken off at 07.15 to fly an artillery observation sortie. They had been working successfully for some little while with the 188th Siege Battery before 15 German fighters turned up around 09.00. Two Fokker Triplanes detached themselves from the group and headed down on the Big Ack, as the AWFK8s were known. Peacock grabbed his rear Lewis gun and began firing at the leading Fokker as it closed in. It was so close he just couldn't miss and despite the German's fire slashing into the two-seater, Peacock held his aim and the enemy fighter staggered away and went down, being seen to crash behind the front lines, right into a Nissen hut. Hughes broke off the sortie in case his machine had been damaged and the two men congratulated themselves on their good fortune. Not only had they survived, they had downed an opponent; they landed at 09.30. Plange had been flying DrI No. 453/17, the wreckage becoming 2nd Brigade's G.10.

Eleven minutes after they had landed, the Sopwith Dolphin-equipped 19 Squadron sent out a squadron patrol of all three Flights, two of A Flight above, five of B Flight in the middle, and six C Flight machines below. They ran into nine German fighters in company with two two-seaters, B and C Flights diving at them, leaving A Flight's two machines above as cover. Major A D Carter DSO, a Canadian of Major rank but not the Squadron CO, although he was a flight commander, went for the two-seaters with another pilot. No sooner had they started the attack than Carter's companion noticed his flight commander start to drop away.

Above them, the others were engaging the German scouts, two being seen to spin away, but two or three of the British pilots also saw the

Dolphin losing height, as if it had suffered engine trouble. However, it didn't take long for a couple of Germans to swoop down and attack the British aircraft, and not long afterwards the machine was seen on the ground, as if it had made a forced landing. Albert Carter did not get back (C4017), and in fact he was claimed by Leutnant Paul Billik, leader of Jasta 52, the Dolphin being his 17th victory. Carter got down all right and was taken prisoner, but he had achieved 28 victories himself. Whether he had suffered engine trouble, had been hit by one of the two-seater gunners, or was simply brought down by Billik is unclear. Billik himself would raise his score to 31 by August, only to suffer a similar fate to Carter, being brought down and taken prisoner, on 10 August.

Downed in a collision

Two aces met, almost literally, over the front lines shortly before midday. Captain H G White of 29 Squadron was leading a patrol out on a balloon hunt and spotted nine German fighters waiting to pounce on them. Hugh White was something of a veteran already, despite his 20 years, for after Sandhurst he had joined the RFC in 1916 and had flown with 20 Squadron for 11 months during 1916–17. After a rest he returned to France as a flight commander with 29 Squadron, flying SE5s and had a personal score of six by 19 May. Hugh White, who rose to be an Air Vice-Marshal, died in 1983, but several years earlier this author had a number of letters from him, including an account of the encounter on this May morning.

Following the action on this day, which will be described below, 29's CO put forward a recommendation for an Immediate Award for his flight commander, but strangely, despite his record and his actions, Hugh White was never decorated for his WWI service. Hugh White told me in correspondence:

> 'The incident occurred when we were shooting up a line of observation balloons and that having seen the nine enemy Pfalz Scouts approaching from above and my chaps having failed to react to my recall/combat red Very Light signal, I decided to engage the nine, in the hopes of keeping them occupied until the rest of my formation followed on and joined in.
>
> Unhappily I collided with one and after blowing him to bits, I made for the lines as best I could, followed closely by first one enemy aircraft, whose gun jammed after two or three shots, and his place then taken by yet another, but as it never occurred to me to do otherwise than attempt to cross our lines, I tried seeing what throwing out some remaining Lewis gun ammunition drums might achieve.
>
> Lo and behold the very first one disintegrated the propellor of the

following aircraft which had just opened fire. After that my Flight had arrived to see the rest of them off and to keep me free from further interference.'

According to his recommendation for an award, the action went like this:

'Captain White was left alone. He dived on one of the enemy aircraft and fired about 100 rounds at very close range. The EA 'zoomed' to the left, and its top plane caught the leading edge of Captain White's machine, causing the EA to turn a cartwheel over his machine. The shock of the collision flung Captain White forward on to the gun mounting, and stopped the engine. The EA went down into a dive and Captain White, expecting his machine to break up at any moment, dived after it, firing about 100 rounds. The right wing of the enemy aircraft fell off, and it went down completely out of control.

Captain White then turned round, and turned west to endeavour to recross the lines. He was followed back by five EA scouts for some distance until these were driven off by friendly machines. Captain White managed to keep his machine fairly straight and level by putting on hard left bank, left rudder, and leaning over the side of the cockpit. Near the ground the machine became uncontrollable and it crashed on landing near Eecke.

The centre section wings of the machine were broken, and the right hand planes had anhedral instead of dihedral. The right-hand top plane was badly damaged, but the main spars held. The fabric was completely torn off.'

The luckless German pilot was Vizefeldwebel Karl Pech, who dived straight into the ground and was killed. By one of those strange quirks of fate, he was from Jasta 29, White with 29 Squadron. Pech had joined his Jasta in January 1918 and had scored nine kills, six during May alone. The Germans recorded his loss at 12.03, and White had been flying SE5a D3942 to the Bailleul area. White received credit for his seventh victory.

The Americans lose Lufbery
There can be few WWI formations that conjur up a more gallant bunch than the Escadrille Lafayette. Formed officially as Escadrille N124 its pilots were, apart from its French CO and flight commanders, all American volunteers who had mostly arrived by first enlisting in the Foreign Legion.

Though the unit's achievements were modest, much has been written of N124 and its Americans, especially of its charismatic top scorer,

Gervais Raoul Lufbery. Actually born in France of an American father and French mother, young Lufbery ran away from home at 17 and travelled the world seeking adventure, even going to his father's home country in 1906 when he was 21. Meeting the French aviator Marc Pourpe in Calcutta six years later he became his mechanic, returning to France with him when war came. When Pourpe was killed, Lufbery decided to become a pilot too. The rest is history.

With something like 16 victories and several more which were never confirmed between 1916–17, Lufbery became a hero in both France and America, decorated with the Medaille Militaire, Légion d'Honneur and Croix de Guerre with ten Palmes, while the British awarded him the Military Medal when he was still an NCO. He then transferred to the American Air Service with the rank of Major, and took command of the 94th US Aero Squadron.

On Sunday 19 May 1918, he took off in a Nieuport 28 to intercept a German two-seater but in the engagement his machine was hit by the observer and set on fire. Rather than burn to death in the air, Lufbery jumped, crashing to earth in a flower garden in the village of Maron, north of Nancy. He was 33. His victors, Unteroffizier Kirschbaum and Leutnant Scheibe of Reihenbildzug Nr.3, were immediately set upon by French fighters and brought down at Mars-la-Tour and taken prisoner.

Lewis Whitehead is lost
On 20 May, Captain Lewis Whitehead of 65 Squadron, who had been in the combat in which Walter Ewers was shot down five days earlier, met his own fate. The Squadron was out on patrol that evening in the Albert region and engaged several Fokker Triplanes. Captain John Gilmour MC, one of the other flight commanders, claimed one Triplane destroyed south-east of Albert, which broke up in the air, but Whitehead, in D1876, failed to return. Vizefeldwebel Fritz Rumey of Jasta 5 got him over Morlancourt, his 21st victory. There is no convenient loss to fit Gilmour's claim.

Paul Baer's brief moment of glory
From Fort Wayne, Indiana, 1st Lieutenant Paul Frank Baer joined the Lafayette Flying Corps and served with the French Spa80 Escadrille over the winter of 1917–18. In the new year he was assigned to the American 103rd Aero and on 11 March beame the first US fighter pilot to shoot down a German aircraft. By mid-May he had scored nine kills plus seven probables and had been awarded with the DSC and Oak Leaf Cluster.

However, in gaining his ninth victory on the morning of 22 May in a fight with Jasta 18, near Armentières he was brought down behind the

APRIL – MAY 1918

lines and taken prisoner. Two of the 103rd were shot down, the claimants for Baer and the other man being Vizefeldwebel Debenitz and Leutnant Hans Müller, the latter's fifth of 12 victories. After release from PoW camp, Baer became a Chevalier of the Légion d'Honneur; he died in a plane crash in Hong Kong Harbour in December 1930.

The 'Fours' knock out Oberlander

Three days after shooting down his sixth victory, Leutnant Hans Oberlander of Jasta 30 was brought down and wounded, which effectively put him out of the fighting war. He had scored one victory in 1916 with an unknown unit, most likely a two-seater Abteilung, but had been serving with Jasta 30 since May 1917.

On the morning of 23 May, de Havilland bombers of 27 Squadron mounted a raid to Carvin, taking off shortly after 06.00. Once over the lines they began to be stalked by eight Albatros Scouts who followed them to the target but made no attempt to intercept them until they had bombed and turned for home. In fact 13 bombers had set out but only eight or nine had made the attack.

The fighters finally attacked, and two bombers, one DH4 and one DH9 were lost, one falling to Leutnant Dieter Collin, CO of Jasta 56 at 09.30 (G), a second apparently losing its engine and crashing. Without doubt the gunner of the one shot down would have returned fire, and two more reported combats; B5506, Captain Doyle/Lt Chester – 600 rounds – and C2086, Lts Coghlan and Dunton – 500 rounds. One of them must have hit Oberlander, who was wounded and was lucky to survive. As it was he was out of action for some time, and although he eventually came back to Jasta 30 in September, he did not add to his score.

A two-seater gets D J Bell

There were few easy targets, and let it be said that attacking a two-seater was every bit as dangerous as a single-seater, sometimes more so. After all, one could, with luck, surprise a single-seater, but however one might sneak up on a two-seater, there was always a good chance of the observer spotting you. As you dived down, an observer with a stout nerve and steady hand was able to take a good aim on an approaching aircraft.

On 27 May, 3 Squadron had sent off Camels to engage an aircraft picked up by wireless interception over the 3rd Army Battle Front. The Camels had taken off at 12.10 and Captain Douglas John Bell MC and Bar, victor in 19 combats, found the offending two-seater and attacked. Two of his companions, Lieutenants Will Hubbard and L A Hamilton, also attacked the Albatros two-seater, but the observer hit Bell's Camel (C6730). The Camel went down and the two Lieutenants watched in horror as the fighter fell to pieces in the air, leaving Bell to plummet to his death into enemy lines. The Albatros was believed to have been crewed by Gefreiter Rosenau and Leutnant Werner Heinzelmann of FAA259 (killed 19 August 1918).

Bob Little falls to a night raider
Captain R A Little DSO and Bar, DSC and Bar, of 203 Squadron scored his 47th victory on 22 May. However, on the night of 27 May he took off in Camel D3416 to intercept night bombers, actually joining combat with a Gotha over his airfield at Filscamp Farm.

He had volunteered to take off at 22.30 and some time later machine-gun fire was heard. Bob Little did not return and land, and it was assumed he had put down elsewhere, for after all there was no radio, radar or lights, with which future night aviators would avail themselves as a matter of routine. However, when by next morning no call had been received from him to say where he was, any uncertainty was immediately ended by a call from 201 Squadron to say that Little's wrecked Camel had been found at Noeux, near Auxi-le-Château. He had been wounded in the thigh by a machine-gun bullet and had crashed, dying from the effects of the wound and injuries sustained in the crash. It transpired that during his attack on the Gotha, a searchlight had picked him up, not only blinding him but giving the German gunner a good view of his Camel. He had been hit in both upper thighs and, following the crash, had bled to death before help could reach him. This high-scoring Australian, from Melbourne, was still some weeks away from his 23rd birthday, and he left a wife at St Margarets, Dover.

CHAPTER TEN

May – June 1918

What happened to Rudolf Windisch?
The month of May continued to be costly for the aces. Rudolf Windisch was about due for a Pour le Mérite having scored over 20 victories by the middle of May, making it 22 on the 27th with a Spad two-seater. All had been scored on the French front over the last nine months, plus a balloon he destroyed when on two-seaters himself.

From Dresden, this 21-year-old had been with Jasta 32b and now commanded Jasta 66. On this 27 May, there was an attack on a French airfield at Courvelles from which Windisch failed to get home. Not that attacks by fighters on any Allied airfield were commonplace, but this one was intercepted by French fighters and Windisch was brought down. The most likely French unit involved was Spa76, who fought a dozen German scouts, place not stated, but it ties in with the airfield location. For some reason the Germans were certain that Windisch had been brought down alive; perhaps one of his pilots had seen him on the ground. In any event, it seemed certain enough for the authorities to continue with his nomination for the Blue Max, something which would not have proceeded had they believed him dead.

Of course, no official word came from him, and at the end of the war he did not return home. So was he in fact killed, or mortally wounded? If this was the case, one would have imagined some sort of joyous report from the French to the effect that they had downed a high-scoring ace; it would certainly have made the headlines. A few stories began to emerge, that he had been shot while trying to escape, even suggesting he had been trying to steal an aeroplane, but everything seemed shrouded in mystery. Perhaps the simple answer is that he was gunned down in the heat of the moment soon after landing, and the French were too embarrassed to admit it.

On the 28th a successful Bristol Fighter pilot, Lieutenant R G Bennett of 20 Squadron, who had with his various observers claimed nine victories since the beginning of the year, was shot down. Bennett had joined the RFC as an equipment officer, only later volunteering for pilot training. He was part of an OP on this morning, taking off at 09.17 with Lieutenant

G T C Salter MC in the back. They were last seen at 10.30 near Neuf Berquin, very close to the spot where Leutnant Kurt Baier of Jasta 18 shot down what in the Nachrichtenblatt is recorded as a DH4. However, it was a BF2b which came down at La Gorgue at 11.20, so time and place match pretty well. It was the German's first and only victory. He served in the Luftwaffe in the 1930s and in WWII, becoming a General-Leutnant in March 1945. He died in 1983.

'Lobo' Benbow
Thursday evening of 30 May saw another veteran air fighter depart. Captain Edwin Louis Benbow MC, despite his 22 years, had seen considerable service. He was in France at the beginning of 1915, aged 19, with the Royal Field Artillery, and a year later was seconded to the RFC as an observer. After pilot training he went to 40 Squadron in 1916 where he flew FE8 pusher scouts, claiming no fewer than eight victories on this type, not noted for its longevity in combat. He even survived fights with Richthofen's Jasta 11, but was wounded on 19 March 1917.

After a rest in England he became a flight commander with 85 Squadron, returning to France in 1918. However, before being able to get back into the swing of things, he was killed. He had led out an evening patrol, at 18.55, flying SE5a C1862, in search of enemy machines and failed to return. He was shot down by Oberleutnant Hans-Eberhardt Gandert, commander of Jasta 51, his third victory, over Nieppe Forest at 20.35(G). He would end the war with eight kills, and like Kurt Baier would become a WWII Luftwaffe General. (Previous references to Kurt Schönfelder as being the victor over Benbow are not correct.)

On the last day of May, an American serving with No.1 Squadron RAF, First Lieutenant Duerson 'Dewey' Knight, from Chicago, was lucky to survive a shoot-down. By this date he had already shared in four victories and would continue to reach a score of ten by August.

Flying SE5a C6479 on an early morning OP along with nine others, he was in action with German fighters over the lines and had his main fuel tank hit. With his engine stopped he dived west from Steenwerck with three Albatros Scouts snapping at his heels, crash-landing 200 yards from the front-line trenches. The SE5 was dashed to pieces but Knight emerged unharmed, thankful for being safe and on the right side of the lines. Unfortunately there is no obvious candidate for his victor. No SE5 machines were claimed, certainly none credited in the Nachrichtenblatt. That he came down so close to the lines (and his aircraft must have been seen wrecked) would normally constitute a victory but perhaps it either wasn't or it was a late confirmation. It didn't affect Knight, however: he helped shoot down a Pfalz the next day.

MAY – JUNE 1918

Walter Böning was lucky too

Bavarian Walter Böning could also count 31 May as a lucky day, despite being wounded. Leutnant Böning had raised his score to 17 the previous day with two Camel kills, and was undoubtedly on his way to a Pour le Mérite. He had been in action since 1914, firstly with a Bavarian infantry regiment then as a fighter pilot with Jasta 9 in 1917 before taking command of Jasta 76b that October. He had been well rewarded for his actions: Iron Cross 1st and 2nd Class, Gold Bravery Medal, Military Merit Order 4th Class with Swords, Crown with Swords 4th Class of the same Order, the Oldenburg Friedrich-August Cross 1st and 2nd Class from his home town, and finally the Knight's Cross of the Royal Hohenzollern House Order, usually a step away from the coveted Blue Max. However, it all came to an end on 31 May.

Jasta 76b was on a late afternoon sortie and spotted a formation of Sopwith Camels south-east of Bapaume. It was nearing 19.00 hrs German time. The Camels were from 70 Squadron and the two formations joined combat. Away to the west, two other pilots of 70 Squadron were also in the air. Captain John Todd MC DFC was a veteran air fighter who had, strangely, also shot down two Albatros Scouts the previous day to bring his score to 12, and had downed number 13 this very morning. He was flying an engine test to the rear of the British lines (C1670), in company with Lieutenant V C Chapman. He saw the Squadron patrol begin the fight with eight Albatros Scouts and could not resist joining in.

Diving in amongst the dog fight he quickly selected one German, put himself on its tail and began firing a burst of 50 rounds from close up (ten yards!). The Albatros burst into flames and went down. As he turned he saw another Albatros spinning earthwards out of control, hit by Lieutenant D H S Gilbertson (D6648). Lieutenant K B Watson had also fired into an Albatros and watched as it fell away and crashed. He then dived on it, firing a burst into the wreck from 400 feet before pulling away. Jasta 76b lost two, although their version differs slightly from 70 Squadron's. As the two formations clashed, Böning had a slight collision with one of his pilots, Vizefeldwebel Georg Markert, who had scored his first and only victory the previous day. Obviously distracted by this event the veteran Staffelführer was at a momentary disadvantage, and in air combat such a distraction often proved fatal.

As he pulled aside from Markert, so he was hit, most probably by the fire from Ken Watson, a 20-year-old from Ontario (he would be 21 on 5 June). Todd, meanwhile, caught the Albatros flown by Markert and set it on fire. Oddly enough, Dennis Gilbertson was not credited with the out of control victory seen by Todd, so it was either not confirmed by Wing or perhaps someone else saw the German pull out and fly off. After all, most German aircraft sent down 'out of control' had pilots that were merely getting out of a tight spot, and once down and clear of a battle they generally levelled out and flew home, not only wiser men but living ones too.

In any event, Jasta 76b had been up against a good unit. Todd was to go on to achieve 18 victories, and Gilbertson had five by September. Although it was Ken Watson's first kill, he too became an ace, downing his ninth victory on 4 November. He returned to Canada with a DFC and lived to see his 63rd year. Böning, however, was out of the war, for in the action he had been seriously wounded in the left thigh. He had lost consciousness due to blood loss when trying to put his Albatros DVa (5765/17) down and crashed. Markert fell in flames near Fricourt, in the area Todd noted. The only problem with this scenario is that Böning is reported to have crashed at his base airfield, so it seems that Todd and Watson claimed the same aircraft. Did Gilbertson, therefore, wound the German ace who then staggered away to the west? Or did Watson wound the German, but watched Markert crash and shoot his machine up? It would be nice to know.

The other ace to fall on this final May day was Leutnant Viktor von Pressentin gen von Rautter of Jasta 4. This former Uhlan Guards Regiment officer, who like so many others had become a fighter pilot via the route of two-seater observer, had achieved an impressive 15 victories since March. However, his 15th and last went down just moments before he did. Jasta 4 were in action against French Bréguet XIV bombers of Escadrille Br29, south-west of Soissons. He and his Staffelführer, Ernst Udet, both downed a Bréguet but then von Pressentin went down, hit and killed by return fire from the others.

Roderic Dallas is killed
Major R S Dallas DSO, DSC and Bar, Croix de Guerre, was one of the great air fighters of WWI. An Australian from Queensland, he was nearing his 27th birthday on 1 June. Having enlisted in the Australian army in 1913 and commissioned, he had decided to transfer to the RFC when war came but was rejected. So he tried the RNAS, was accepted, and the rest is history.

Flying with 1 Naval Wing, then 1 Naval Squadron, he had been an exponent of the Sopwith Triplane and most of his first 20 victories were scored while flying them. With three more on Camels, he had then been given command of 40 Squadron RAF, equipped with SE5s. Despite having flown rotary-engined Nieuports and Sopwith fighters, he adapted well to the in-line engine of the SE5a and had brought his score to over 30 by the end of May 1918. As a squadron commander, he was not obliged to fly combat missions. At various stages of the war, such men were even ordered not to fly, in case they and their experience were lost. This was sometimes ignored by the more forthright and active air fighters. By 1918, squadron commanders did fly, often because they wanted to, and felt the need to lead from the front. Sometimes, however, they did not want to upset the Flight patrol system, and so either accompanied the patrol as just a patrol member, or simply flew off on their own. However, this latter was always dangerous; the day of the lone air fighter was now past.

On Saturday 1 June, Dallas flew as part of a bombing OP of eight aircraft, led out by Captain C O Rushden, to the Merville area at 05.30. One SE had to abort but the others bombed and strafed German positions and were back by 07.30. At 10.10, Dallas climbed into his SE5 again (D3530) and took off on a lone patrol. Obviously too experienced to fly into enemy air space alone, he poodled along on the British side, probably waiting for some two-seater to come across, or perhaps watching for a fighter or two to dart at the balloon lines. However, the hunter was being hunted. He had obviously been spotted for a couple of Fokker Triplanes of Jasta 14 came across the lines, presumably unseen by the Australian, and came at him from the west. As Dallas was undoubtedly still looking eastwards for any sign of German air activity, the first he probably knew of the Triplanes' presence was when the leading one opened fire on him. In the leading Fokker sat Jasta 14's Staffelführer, Leutnant Johannes Werner, who had started out to make an attack on British balloons, but had spotted the lone SE5 and was able to make a surprise attack. Dallas went down under fire from 'Hans' Werner's Tripehound and crashed near Lieven. The German had scored his sixth victory, timed at 12.35(G). He gained just one more although he remained in command of Jasta 14 for the rest of the war.

This same June afternoon, Captain W J Cairnes of 74 Squadron was also killed. Bill Cairnes, an Irishman, was 22 and had seen action with 19 Squadron in 1917, and scored three victories. Now as a flight commander with 74 Squadron he had raised his score to six but during an OP, flying SE5 C6443, he was last seen in a dog fight north-east of Estaires at 16.25. At this time ground observers saw an SE5 break up after being engaged by enemy aircraft.

He was shot down by Leutnant Paul Billik of Jasta 52, who had downed more than one British ace (Malone and Carter) and would down at least one more in his score of 31. He had also killed Major Leonard Tilney MC, the CO of 40 Squadron, whose death led to Dallas being made CO of that Squadron.

On 2 June, Lieutenant P M Dennett of 208 Squadron was killed. Pruett

Dennent, from Southsea, Hampshire, was only 19 but had seen action in late 1917 when his squadron was still 8 Naval RNAS. Seven victories and the experience that went with them did not help him survive a late morning OP with five companions.

They had a fight with several Albatros Scouts over Estaires, although he was still with the patrol when this fight ended, but he then disappeared and failed to return. His Camel, D1854, does not appear to have been claimed by anyone. In the past he has been credited to Oberflugmeister Kurt Schönfelder of Jasta 7 (a navy pilot flying with this army unit), but his claim was timed at 20.55(G), and was probably a 54 Squadron machine. Had Dennett perhaps been wounded in the fight and crashed later, or had he gone down to strafe a ground target and been hit by ground fire? His end remains a mystery and he has no known grave.

Aces from Spa154 and Spa99
Maréchal-des-Logis Xavier Jean-Marie Louis Moissinac was 21 and a former artillery man prior to joining the French air service. Joining Spa154 in the late summer of 1917 had been a struggle, for he had been retained as an instructor, and few pilots welcomed a delay in joining a front-line escadrille. By the end of May he had scored seven victories, three on 30 May, shared with other pilots.

On 3 June he was flying a Spad XIII north of Dormans, about 20 kilometres east-north-east of Château Thierry on the Marne. The French patrol ran into enemy fighters and following the dog fight, Moissinac was found to be missing. Three German pilots made claims over Spads, just to the north-west of Château Thierry; Karl Bolle and Hermann Frommherz of Jasta Boelcke (victories 13 and three), and Erich Löwenhardt of Jasta 10 for his 26th. Unfortunately the time of the Frenchman's loss is unclear.

On the 4th Capitaine Jean Marie Emile Derode, commander of Spa99 failed to return from patrol. He had already received the Légion d'Honneur and the Croix de Guerre with eight Palmes (plus the Belgian Croix de Guerre) but his ninth Palme was his last. The citation for this noted a victory on 4 June during a fight with superior numbers of the enemy, and he too was flying in the Château Thierry area. It is unfortunate again that a time is not known, but there were several Spad claims in the general area, including two by Jasta 66, one by Werner Preuss, another by Willi Schulz. For Schulz it was his fifth victory, Preuss his first – of an eventual 22.

Leutnant Karl Albert Mendel of Jasta 18 was shot down on 6 June, two days following his seventh victory. A patrol of nine Pfalz Scouts, noted as being silver doped with red markings on the wings, was engaged on this evening, near Estaires. The British patrol came from 29 Squadron, led by Captain R C L Holme MC. It was 17.45 (Allied time), the SEs flying at 13,000 feet, as five of the nine German fighters detached themselves from the others and dived. Lieutenant B R Rolfe (D5996) got on to the tail of one, firing 150 rounds from close range, following the German down to 9,000 feet. The machine began to leave a trail of smoke followed by flames, Holme and Lieutenant H A Whittaker seeing it falling on fire and completely out of control.

Meantime, Lieutenants A E Reed (D5970) and C G Ross (D5963) fought another Pfalz down from 13,000 to 1,000 feet, both firing into the scout from close range, and Lieutenant T S Harrison saw it crash northwest of Estaires at about 18.00 hrs. Arthur Reed and Charles Ross, both South Africans, would both end the war as high scoring aces, 19 and 20 victories respectively. For Rolfe this was the first of three. Jasta 18 lost two pilots in this action, Mendel and Leutnant Hans Schultz. However, Schultz was brought down near Hazebrouck in one of the relatively new Fokker DVIIs (368/18), inside Allied lines (2 Brigade G.14), by Lieutenant C H R Lagesse.

Fritz Rumey's 23rd victory
Fritz Rumey, Otto Könnecke and Josef Mai were Jasta 5's three high-scoring NCO pilots, who would score 45, 35 and 30 victories respectively. Only Rumey would not survive the conflict, but all three would receive commissions and the Blue Max.

On 7 June 1918 Captains I D R McDonald MC and C N Lowe MC led an OP to Rosières, taking off at 10.45, with Lieutenants Dawe, Southey, Crossen and Foster, all experienced fliers. Two formations of seven and eight Fokker Triplanes were encountered an hour later, one being above the SE5s, so McDonald manoeuvred for a better position. James Dawe, however, who had gained his eighth victory four days earlier, immediately dived on one of the lowest enemy machines and two Triplanes at once got on his tail. McDonald attacked, firing 20 rounds, but had gun trouble. Cyril Lowe tried to help but he also had problems with his guns, but clearing them, he dived on a fighter firing 20 rounds at point-blank range, setting the German on fire. Roy McDonald had also cleared his guns so attacked the second Triplane firing 100 rounds at close range. The Fokker's top right-hand wing folded back and it went into cloud, into which Dawe had also disappeared, the latter being seen shortly afterwards diving east, alone, at 7,000 feet. Southey had also sent down a Triplane out of control and so had Crossen.

Crossen and another pilot, which the former took to be Dawe, then dived on two Triplanes which promptly dived east, and on looking round Crossen found himself alone. Lieutenant George Foster (Canadian) also went down on a Triplane, firing 60 rounds at close range, and trailing some smoke it spun upside down into a field north of Rosières and caught fire. Thus the SE5 pilots remained in control, having destroyed three, probably another and damaged a fifth. These machines, it was noted, had bright red tails with a zig-zag marking along the top of their fuselages.

It seems that the two German units were Jasta 5 and 19, but whereas Fritz Rumey was credited with an SE5 down north of Rosières, this followed a disputed claim by Jasta 19's Leutnant Arthur Rahn, which if confirmed would have been his seventh victory. (He was wounded in the hand on 17 July so never achieved victory number seven.) Jasta 19 lost one pilot in this fight, who came down the wrong side of the lines and was taken prisoner. Jasta 5 suffered no pilot losses. Nineteen-year-old Dawe, from Rickmansworth, in B611, was killed.

Lieutenant Alexandre Paul Leon Madeleine Marty, from Toulon, a former cavalry man and victor in six combats flying with Spa77 and then Spa90, fell on 9 June. Holder of the Légion d'Honneur, he most probably fell to ground fire, Flakzug 118 claiming a Spad XIII. He came down near Plainfaing at 09.00; other claims this day were by a German two-seater and two more to ground fire.

Captain Walter Tyrrell MC
One of three soldier sons of Alderman Tyrrell of Belfast, Walter had joined the RFC in 1917 and flown with 32 Squadron, flying DH5s and later SE5s. By June 1918 he had 17 victories and a Military Cross.

On 9 June 1918, a further German offensive had opened, which became known as the Battle of Matz, which was down on the French Front, but which was supported by British troops and British RAF squadrons. The latter came mostly in the tried and tested ground attack sorties, 32 Squadron operating along the front between Montdidier and Lassigny on the first sortie of the day, taking off at 06.00. Near Maignelay at 07.00, Tyrrell's SE5 (B8391) was seen to dive into the ground from 1,000 feet. The machine caught fire and Tyrrell was burnt to death.

It has to be assumed that he had been hit by ground fire, managed to climb to 1,000 feet before passing out, or perhaps dying, and it certainly looked as if the machine quite suddenly became uncontrolled. There is no mention of enemy fighters around, although some sources give Leutnant Fritz Hencke of Jasta 13 this SE5 as a victory, shot down near Maignelay. No.32 Squadron did have a pilot wounded in a dawn ground attack sortie, but this was definitely ground fire, and the only other SE5 casualty was Lieutenant F R G McCall MC of 41 Squadron. He shot down a two-seater at 06.25 and was then shot up in the engine by another (?) enemy aircraft at 07.30 near Berry but force-landed on the right side of the lines without injury to himself. Was this a result of Hencke's fire? If it was, he nearly stopped the scoring run of a man destined to become one of the top Canadian aces of the war. Fred McCall, from Vernon, British Columbia, had seven victories, counting the two-seater this morning. He would go on to achieve 35 by mid-August.

Bruno Loerzer's brother Fritz
By 1918, Hauptmann Bruno Loerzer had the Pour le Mérite and following the command of Jasta 26 was now leading Jagdgeschwader Nr.III. At this time he had scored 25 of his eventual 44 victories. (When flying two-seaters earlier in the war, his observer had been Hermann Göring.)

His brother Fritz had seen service with Jasta 6 and briefly as CO of Jasta 63, but on forming JGIII Bruno had had him take command of his old Jasta 26. By the end of May, 24-year-old Fritz had downed 11 opponents. On 12 June Loerzer went down and crash-landed inside French lines, due, he said later, to the rudder controls of his Fokker DVII failing during combat. He came to earth near Cutry, seven miles south-west of Soissons. What is not clear is whether his controls were damaged by enemy fire. There does not appear to be any claimant, especially as he ended up as a prisoner, so it might just have been control failure pure and simple. In the past it has been recorded that he may have been the victim of Captain Roy Phillips of 2 AFC Squadron, who claimed two Fokker Triplanes, an LVG and a Fokker DVII around noon this day. However, ruling out the Triplanes, which in any event were shot down between Noyon and Montdidier, and may well have been two Jasta 69 pilots, the DVII crashed near Gurney, in a scrap with 43 Squadron and 2 AFC. This fight, which included Jasta 15, was around Compiègne, still some way west of Cutry.

Jim Belgrave's last combat
Captain James Dacres Belgrave MC was another experienced air fighter the RAF could ill afford to lose. He had flown Sopwith two-seaters with 45 Squadron in 1917, flown Pups and SE5s in Home Defence and finally become a flight commander with 60 Squadron in April 1918. By the morning of 13 June he had scored 17 victories.

Soon after dawn on 13 June, he took off leading his Flight and over the front ran into a two-seater. They shot it down to the east of Albert, but not before the observer had scored hits on two of the SEs. Lieutenant R G Lewis (C9498) went down to a forced landing, ending up as a prisoner, but Belgrave (D9588) dived away and was not seen again. He was still only 21 years old. A two-seater crew of Unteroffizier Joern and Leutnant Clauditz were credited with one SE this date, their first victory.

Ernest Kelly, 1 Squadron
Second Lieutenant E T S Kelly, from Picton, Ontario, was one week short of his 20th birthday on this same 13 June. Captain P J Clayson MC led an OP to the Langemarck–Passchendaele area at 20.25 hrs and ran into a bunch of red and yellow Pfalz Scouts. Clayson attacked one head-on which went down pouring smoke, then burst into flames near the ground.

However, when the fight broke up, Ernest Kelly did not re-form with the others, last seen near Laventie in B8508, at 20.50. The problem with that statement is Laventie is 34 kilometres south-west of the patrol area. Previously it has been suggested that he had been shot down by Leutnant Otto Creutzmann of Jasta 43, who claimed a 'Sopwith' south-west of Aubers, but the time given is 09.50 in the morning (unless this is a mistake for 21.50, i.e. 20.50 British time). Admittedly there are no Sopwith losses this day, but Aubers is 35 kilometres south-west of the patrol area, although only five kilometres south-east of Laventie. Why would Kelly have flown so far south? Was he chased? With no other obvious candidates, perhaps Creutzmann did get him and the time has been noted incorrectly. Whatever the truth, Creutzmann was now given command of Jasta 46, having now achieved five victories. He ended the war with eight. And there is no help in trying to tie up Clayson's flamer. There are no records that Jasta 43 or any other unit in 1 Squadron's combat area lost a pilot.

The other Schäfer
Jasta 11's famous ace, Karl Emil Schäfer had been killed in 1917, but another Schäfer, Leutnant Hugo, flew with Jasta 15, and this month would see his 24th birthday. He had enlisted in 1915, was commissioned and joined the Air Service. By the end of 1917 he was with Jasta 15 and by mid-June had scored five victories.

On 17 June, 24 Squadron flew a midday patrol to the Alaincourt-Hangest area and ran into Jasta 15 with their new Fokker DVIIs which were making life difficult for the DH4s of 27 Squadron. One of the SE5 pilots was the American Lieutenant W C Lambert, who had started the day with his score at seven. The Ohio-born 23-year-old was attacked by two enemy fighters which were in turn engaged by Lieutenant W E Selwyn, leaving Lambert to fight with just one. They had a long fight but eventually Lambert's fire smashed the tailplane and the Fokker went down in a spin. Lambert was then engaged with another Fokker but he fired 200 rounds into it and the fighter began to spew blue smoke before going down in a spinning nose-dive staining the sky with more smoke as it did so. One of these appears to have been Schäfer, probably the second one, which had just shot down a DH4. Schäfer managed to make a forced landing behind his lines, safe, but his machine was a write-off. However, he survived and ended the war with 11 victories.

No.24 Squadron gets Wüsthoff
The day after Hugo Schäfer was nearly knocked out, 24 Squadron scored a more notable victory, or at least a much more publicised one. Leutnant Kurt Wüsthoff was a holder of the Pour le Mérite and had 27 victories, all but one scored in 1917 with Jasta 4. He became Staffelführer on 22 February 1918 but had only added one more victory to his score sheet before becoming a staff officer with JGI in March. He had then been with FEA 13 for a short period while officially on sick leave but was then given command of Jasta 15 within JGIII on 16 June. He now found himself, like so many others, back at the front, rusty at being away from the action and therefore vulnerable. Just how vulnerable he discovered the next day.

It is one of those strange things but he is remembered more for the aeroplane he was flying than his own achievements. Jasta 15 was equipped with the new Fokker DVII biplane that was fast becoming the most successful of the German fighters in WWI. However, he did not yet have his very own machine, so on the 17th he borrowed the machine normally flown by Georg Hantelmann (382/18). As this machine more or less survived the day's encounter and was well photographed, it probably became one of the most well known machines, featured in many journals and later in craft magazines. It was, of course, very distinctive. It had lozenge fabric wings, and Jasta 15's Prussian blue fuselage, red nose, white tail fin, and Hantelmann's usual personal insignia of a huge white skull and crossbones on the fuselage sides. Not that Hantelmann had had much chance of flying it himself. (He had only three victories by this time but would end the war with 25, including three aces.)

The SE5s had met ten German fighters and from the reports it does seem as if Wüsthoff was separated from his men and then set upon by several of the British pilots. Lieutenant H D Barton (C6481) fired 200 rounds into the Fokker and it began to go down. Captain Roy McDonald (D3444) also attacked it, then Lieutenant G O Johnson (C1084), firing 20 and 400 rounds respectively. The German had virtually no way out, being surrounded by these men plus Bill Lambert, Palmer, J H Southey, Selwyn, E B Wilson and E W Lindeberg, almost all experienced air fighters.

Wüsthoff force landed near Cachy and turned over, with wounds to both legs, apparently inflicted by an attack by J H Southey. Even a Dolphin of 23 Squadron had got in a burst. That was the end of Kurt Wüsthoff's war. He was to die from injuries sustained in a plane crash in July 1926.

Frank Baylies of the Storks
Massachusetts-born Frank Leamon Baylies died aged 22 on 17 June. He had volunteered for the Ambulance Service in 1916, seeing service on the Somme, at Verdun and in the Argonne, and later went to Serbia. He won the Croix de Guerre for evacuating wounded under fire. Enlisting into French aviation he became a pilot with Spa 73 but then went to the famous Storks (Cigognes) Groupe de Chasse 12, and to the most famous of the four escadrilles, Spa3.

With this unit he claimed 12 victories until he too fell in combat, with Triplanes of Jasta 19. Spa 3 lost two Spads this day, Baylies and Maréchal-des-Logis André Dubonnet, who had four victories by this time, but would score two more in August. Jasta 19 claimed one Spad on the German side, one on the Allied, both west of Roye. Leutnant Rudolf Rienau claimed on the German side, which must have been Baylies. It was the German's second of the six victories he would achieve by the war's end.

Leutnant Wilhelm Schulz had scored six victories with three different Jastas, 16, 41 and 66, between August 1917 and June 1918. He was shot down in flames over Dormans Park on the 25th, probably by two Spa57 pilots, Sous Lieutenants Leon Nuville and Marius Hasdenteufel. Their victory over a Fokker DVII was timed at 18.10, it being Nuville's seventh of 12 victories, Hasdenteufel's fifth and final one; he was killed the next day in an accident.

Kurt Schönfelder's last combat
Twenty-three-year-old Kurt Schönfelder, from Totschem, had learnt to fly before the war, his licence being dated December 1913. Despite being a Naval pilot, he had flown with success with Jasta 7 since early 1917, and now, a year later, had achieved 13 victories, the last being a Camel of 210 Squadron on 21 June. Now five days later (26th) 210 Squadron were to exact their revenge but he had no way of knowing it.

Jasta 7 and 210 Squadron's Camels met west of Armentières around 19.15, the Germans being led by Josef Jacobs, 210 by Captain L P Coombes, a 19-year-old former RNAS pilot who had joined his unit in January while still only 18. To date he had claimed eight victories. Among the men flying with him this evening were two tyro pilots, 18-year-old Ivan Sanderson, from Buckinghamshire, with one victory, and 20-year-old Ken Unger, an American from Newark, New Jersey who had yet to score. Both would become aces with 11 and 14 victories respectively. Coombes (D3387) led the attack on the six hostile scouts, getting on the tail of a Fokker biplane, opening fire, then letting both Sanderson (B7153) and Unger (D9608) fire into it. In all they fired some 300 rounds at the black DVII and it went down in a slow spin, then a series of stalls and dives until finally it crashed about a mile west of the town.

An hour after this fight, Lieutenant E C Eaton from Montreal, flying with 65 Squadron, went down to Fritz Rumey, the German's 25th victory of the war. Edward Eaton had five victories since the previous autumn.

The Camels, led by Captain A A Leitch MC, had come across eight or nine Albatros and Pfalz Scouts while on an OP 11,000 feet east of Albert, at 20.35. The Germans were coming north over the southern part of Morcourt, Leitch climbing to try and gain height as they moved north of Averluy Wood, but the Germans – Jasta 5 – were no fools and kept themselves just that bit higher than the Camels. White AA bursts (British; the German AA fire was black) attracted Leitch's attention about half a mile behind and below, so as they were not gaining on the upper machines, he turned to see what the AA were firing at. This was just what Jasta 5 were waiting for and the German pilots swiftly made a diving turn after the Camels. However, Alfred Leitch, from Manitoba, was no fool and would soon add the DFC to his MC, having gained five victories thus far. Spotting the Germans begin to commit themselves, he turned his Camels back to face the coming attack. One German immediately went down leaving a trail of smoke as Leitch opened up, he then attacked a green-tailed Pfalz, zooming up to fire a short burst and watched as it fell away, its propeller hardly turning. He had to leave it to watch out for others, but later, a British AA battery confirmed one Pfalz crashing near Albert.

Another pilot, the American F E Kindley (D1878) saw a Pfalz close to a Camel's tail, presumably Eaton (D6630), as he had left the formation away to Kindley's right. Field Kindley turned and dived on the Pfalz, also seeing the Camel in a steep right-hand turn, smoke pouring from its engine. The Pfalz turned east, Kindley firing at it but then it half-rolled and went down in a flat spin, on its back, leaving a smoke trail, but Kindley too had to break away due to other enemy machines in the vicinity. Landing back, Leitch claimed two, one crashed and one out of control, while Kindley too was credited with a crashed Pfalz, his first victory. He would go on to score 11 more with the 148th US Squadron. Leitch brought his score to seven but did not score again.

Eaton was killed, claimed and credited to Fritz Rumey. Looking at the combat, it would seem that it was Rumey whom Kindley had seen shoot down the Camel and then being attacked; he had simply half-rolled on to his back and allowed his Pfalz to spin away to a lower altitude, where he would have righted himself and flown home. Jasta 5 did, however, lose a Pfalz and a pilot: Leutnant Wilhelm Lehmann, the Staffelführer, went down over Albert and was killed. He had been with the Jasta since the previous summer, taking command on 18 May 1918. He had four kills.

JGI lose Steinhauser
Leutnant Werner Steinhauser flew with Jasta 11, part of JGI. He had gone the usual route: two-seaters in 1917 (one victory), then on to fighters, joining Jasta 11 in December. Scoring steadily, he now had ten victories.

Jasta 11 had moved down to the French Front to help with the offensive there and had been scoring well. On the morning of the 26th, Jasta 11 flew behind the front and were engaged by Spads south-west of Soissons. Steinhauser was hit and shot down over Neuilly at 08.00. The French had only two confirmed victories this day: Sous Lieutenant Robin of Spa93 claimed his first, a fighter north of Longpont, while two pilots of Spa124, Sous Lieutenant Douillet and Sergent Saimer, shared an enemy machine at Roissey-en-France. More likely was the Spa93 claim.

Two days later the Germans evened the score by downing the French ace Adjutant René Montrion of Spa48, who had 11 victories and eight probables. A Parisian, Montrion had been in action for over a year and had received the Medaille Militaire following his sixth victory. He also had the Croix de Guerre with seven Palmes and one étoile de vermeil and had been nominated to be a Chevalier de la Légion d'Honneur. He had flown out to make a balloon attack near Villers-Hélen and had been shot down by a German fighter. Unfortunately there is again no confirmation of time, so it is difficult to tie up any of the several Spad claims. Among those who did claim in the area were Karl Bolle, Jasta Boelcke, Fritz Friedrichs of Jasta 10 and three pilots of Jasta 4.

On the last day of June, Leutnant Hans Viebig of Jasta 57 was shot down and severely wounded in combat with Camels of 70 Squadron and the SE5s of 41. He had scored his sixth victory during this month. He made a forced landing near Harbonnières, east of Villers-Bretonneux, and did not see further action in the war. With 11 victories claimed by these two RAF units this day, it is anyone's guess as to who scored hits on Viebig's machine. The main combat was south of Bray, above the Somme at around 18.10.

CHAPTER ELEVEN

July 1918

The war in the air was still changing. Certainly it was a very different war to 1917, even to the last spring of 1918. The Germans had failed to achieve a major breakthrough on the Western Front and their High Command knew the vast American effort would soon be coming. Already more and more American airmen were being attached to British squadrons for experience, and the first American combat squadrons were fighting down on the French sectors, and being reinforced all the time.

The Mad Major
The Commanding Officer of 87 Squadron was Major Joseph Creuss Callaghan MC, aged 25. Born in Ireland, he had spent the pre-war years in Texas but returned to the colours in 1914 and joined the Royal Munster Fusiliers before transferring to the RFC the following September. He had already flown a tour with 18 Squadron on FE2s, then became a fighting instructor at Turnberry. It was while at Ayr that he became known as the Mad Major by the locals due to his amazing flying stunts around the neighbourhood. It was all part of the Fighting School's programme, encouraging the pupils destined for fighters to expand their knowledge and expertise in aerobatics and manoeuvring, and it seemed to suit Callaghan well. In London, prior to his going back to France, he became famous for his somewhat robust antics, especially with some of the Americans who were with 85 Squadron, something of a sister unit, so the nickname stuck.

Taking 87 Squadron to France shortly after the German March offensive died down – delayed because most of their Dolphins had been given to another squadron already in France to replace battle losses – Callaghan had added four victories to the one he had claimed with 18 Squadron two years earlier. As mentioned previously, squadron commanders were not required to fly on operational sorties, although many did, and Callaghan was yet another who was not going to sit on his hands. On Tuesday, 2 July 1918, Callaghan took off alone to fly a patrol over the army front in D6371, take-off time noted as 10.15. Over the front he spotted a gaggle of enemy fighters over the other side of the lines and headed for them. Some estimate the number of Germans to have been about 25, but the number didn't worry Callaghan. In the event, it would be an exaggeration to say he attacked all of them, as in reality he probably tried to engage just a few who may have become detached from the main bunch. Anyway, some SE5 pilots some way off spotted the lone Dolphin ploughing into the middle of several German machines to start a fight. Before they could help out, however, they saw the British machine go down in flames. Callaghan had taken on elements of JGII and was shot down by the leader of Jasta 13, Leutnant Franz Büchner. Callaghan went down over Contay at 10.45 British time, the German's seventh victory. He would achieve 40 victories by the end of the war.

Büchner downed another ace five days later (although it could just as easily have been Ulrich Neckel, also of Jasta 13). A Camel patrol of 209 Squadron was on a Special Mission to intercept a German wireless machine, taking off at 10.00, and nineteen minutes later the radio calls stopped as the two-seater was engaged over Morcourt. However, Jasta 13 were in the area as cover and eight fighters came down on the Camels, the leading pilot picking out one which went into a spiral dive from 1,500 feet and crashed into the ground. When the surviving Camels flew home they discovered another of their number was missing.

Two of the Camel pilots had been in the fight which saw the end of Manfred von Richthofen back in April, M S Taylor (D3329) and W R May (C194). Sammy Taylor, from Regina, Saskatchewan, had scored seven victories, while May too had achieved acedom with five (war total 13). Taylor and Lieutenant D Y Hunter were both shot down in this action, credited to Büchner, his eighth, and Neckel, his 20th. Despite his score, Neckel would have to wait some time for his Pour le Mérite.

Unconfirmed victory
Vizefeldwebel Otto Rosenfeld of Jasta 41 was 26. Flying with Jasta 12, then Jasta 41, he had scored 13 victories by July 1918. He could just as easily have sat out the war in a prison camp, having been brought down by ground fire during a balloon attack on 29 December 1917. He got the balloon, his eighth victory, but was 'in the bag'. However, he managed to escape in April and by May was back with Jasta 41.

On 7 July, having downed a Nieuport 28 for his 13th victory, he was hit by fire from another and fell to his death. The Nieuports, apparently, had been attacking the Jasta's airfield at Coincy, where they had arrived two days earlier. Lieutenant Sumner Sewell of the 95th Aero claimed a fighter shot down over Coincy on this day, but did not receive credit for what might have been his third victory. Sewell, later a Captain DSC, ended the war with seven victories. The 95th also lost one machine and pilot (PoW).

No.32 Squadron lose a flight commander

Monday 8 July saw the loss of Captain Arthur Claydon DFC, of 32 Squadron. He was one of the first recipients of the RAF's two new decorations, the Distinguished Flying Cross (for officers) and the Distinguished Flying Medal (for NCOs). (There was also the Air Force Cross and Air Force Medal for non-operational achievements with the air force.) Claydon was a Canadian, and had been with 32 since the previous autumn. Now, with seven victories and a Flight to run, he was an experienced air fighter. On this July day he led a patrol out to the Lille-Carvin area at 06.30, in C1089, also giving cover to Bristol Fighters patrolling below. Between Seclin and Carvin, at 08.15, Clayden saw a fight developing below between Camels and Bristols (73 and 62 Squadrons), and several Fokker DVIIs and Triplanes. Claydon led his large force down – it was a patrol of squadron strength – and joined combat.

Lieutenant A A Callender (C1903), an American from New Orleans, was in the pack. A future ace, he was about to claim his third victory. He attacked a Fokker biplane, firing 75 rounds at it and saw it spin down out of control, but he was unable to watch too long as he was watching Claydon, who was also going down, seemingly under control but without his engine going. He lost sight of him at 4–5,000 feet.

The opposition was Jasta 52, commanded by Paul Billik. Billik was already a highly experienced fighter pilot as mentioned earlier. His score now had reached 22, and in this fight he not only knocked out Claydon but also a Bristol Fighter. No.32 Squadron also lost another pilot and Jasta 52 lost one to the joint efforts of 62 and 73 Squadrons, the latter also losing a Camel.

Jasta 40's Staffelführer falls

The German fighter units were quite different from their RAF counterparts. For one thing they were smaller in size, rarely having no more than ten or a dozen pilots on strength, as opposed to around 18 or more in a British squadron. Whereas the RAF unit was based on three Flights, plus a squadron commander, the Jasta just had its leader, although that depended on whether this was a 'flying' leader or someone who had achieved the position through rank. The rank system was paramount in the German armed forces, and even some highly successful fighter pilots continued to be NCOs for long periods, sometimes never being commissioned. There was also a vast difference between regular and reserve (war duration) commissions. While reserve officers could and did command units, more often choice went to officers with permanent ranks, whether they were a successful pilot or not. Some Jastas had leaders who had scored no victories, or just one or two, while among the Staffel could be two or more high scorers. Where, of course, a big scorer was also the Staffelführer, that Jasta could be a little more motivated. Jasta 11 under von Richthofen was a case in point.

Rank was also not automatic in the German service. One was either an NCO or a second lieutenant – Leutnant. A few became a first lieutenant – Oberleutnant – but it was not an automatic promotion to higher rank if one commanded a unit. In the RAF a flight commander would always be of Captain rank (although perhaps noted as Temporary until confirmed), while a CO was always a Major. In comparison, very few Jasta commanders rose to the dizzy heights of Hauptmann (Captain) and many saw out the war as a mere Leutnant despite long periods in command.

Jasta 40's commander was Leutnant Helmut Dilthey, who came from the Düsseldorf region. In the air service since 1914 he had seen action on the Russian Front in two-seaters before taking a single-seater into combat, while still an NCO. Commissioned and then trained as a fighter pilot, he had joined Jasta 27 in 1917, under Hermann Göring. By early 1918 he had scored six kills and in May was given command of Jasta 40s. A period of leave was taken during the second half of May, then he began leading his unit in earnest, but only scored one more victory, in June. Most of the unit's scoring was done by Carl Degelow, who was destined to succeed Dilthey.

Dilthey met his end on 9 July. DH9 bombers of 107 Squadron flew a bomb raid this morning and were engaged by Jasta 40 on their way to Lille railway station. Dilthey, flying his green and white ringed Albatros DV, closed in as German AA fire began to explode around the aircraft. German sources believe Dilthey was hit by this AA fire and burst into flames, but the RAF record a different ending. In DH9 D1734 were Second Lieutenant J R Brown and AM J P Hazell. Hazell had watched three Germans approach and had begun firing at the leading fighter. After 40 rounds he saw it burst into flames and dive for the ground. Whether hit by AA, Hazell, or both, the 24-year-old fell to his death.

The French lose four

On 16 July the French lost Sous Lieutenant André Jean Louis Barcat, of Spa153, a five-victory ace who would be made a Chevalier de la Légion d'Honneur – posthumously. Barcat was yet another Parisian and former artillery man, who had not moved to the air service until 1917.

Spa153, on patrol in the IV Armée Sector, tangled with Jasta 19 over Suippes, near Malmy and lost two pilots, one being Barcat. It is understood he went down in flames under the guns of Leutnant Hans Pippart, the Staffelführer, for his 17th victory. Spa103 also lost one of their aces this day, Adjutant August Baux, reported to have scored five victories.

Baux, serving in the infantry at the start of the war, was actually taken prisoner in October 1914 but had escaped and then joined the air service. He had received the Medaille Militaire and Croix de Guerre with four Palmes. The Germans claimed at least 18 Spad fighters this day, making it difficult to say which one was Baux.

This is the same story for Adjutant Maurice Joseph Emile Robert of Spa 92 on 19 July, again shot down and killed on the IV Armée Sector. He had five kills and two probables. The fourth Frenchman to fall this week was Sergent André Louis Bosson of Spa 62, on the 20th, another who had joined up in 1914 and risen through the ranks prior to pilot training. His seven victories had brought him the Medaille Militaire and the Croix de Guerre with five Palmes. He probably went down to a JGIII pilot, perhaps Bruno Loerzer or Erich Buder of Jasta 26.

The Germans lose two

Down on the French Front the Germans lost two aces on 18 July, Leutnants Konrad Schwartz of Jasta 66 and Moritz-Waldemar Bretschneider-Bodemer of Jasta 6. Schwartz had downed his fifth victory the previous day, but only his first as commander of Jasta 66, having previously served with Jasta 22. Brought down and taken prisoner, he died of his wounds on 24 August. Bretschneider-Bodemer went down over Grand-Rozoy during a running fight with Bréguet bombers that morning. However, he may also have been shot down by the French ace Lieutenant Georges Felix Madon of Spa38, who claimed two German fighters this date, to bring his score to 38 (of 41).

Offizierstellvertreter Otto Esswein of Jasta 26 had a lucky escape on 16 July. Attacking a balloon his aircraft was set on fire but he took to his parachute and landed safely. Five days later, however, his luck ran out. This 12-victory ace was shot down over Hartennes and was killed.

The only Indian ace

The number of Indian pilots to fly scouts in France can be numbered on the fingers of one hand. The most successful of these was Lieutenant Indra Lal Roy. He came from Calcutta, but was actually at a school in England when the war began. Joining 56 Squadron in 1917 he did not get off to a good start and crashed his SE5a, thus being sent back to England for further training. He was then deemed medically unfit for flying but this was later reversed and he found himself back in France, posted to 40 Squadron in June 1918. In July he claimed five aircraft destroyed and five out of control for which he was recommended for the DFC. This was approved, but before he received it he was killed.

On 22 July, on an early morning OP to Carvin, the SE5 patrol became embroiled in a fight with a number of Fokker DVIIs. Roy, in B180, was seen to fall in flames and crash by a British AA battery at 08.50 during this combat, going down near Carvin. He is sometimes noted as being shot down by Oberleutnant Harald Auffarth, leader of Jasta 29, his 15th victory – an SE5a over Loos (-en-Gohelle, north of Lens, and eight kilometres west of Carvin) at 09.40 German time. This suggests Roy fell inside British lines, therefore it is incorrect to record Auffarth's victory as plain Loos, which suggests it is Loos by Lille, some distance to the north-east and well inside German lines.

However, Roy was buried at Estavelles, inside German lines, which would indicate he fell to someone else. In fact, Jasta 29 claimed two more SE5s in this combat, although no more were lost. However, the American Lieutenant Reed Landis, who claimed a Fokker over Carvin, may have been attacked and was seen to fly low over the lines, or it may have been another SE5 pilot altogether. With no other SE5s down at this time or place, did Auffarth attack an SE5 which in the event got home? Vizefeldwebel Karl Gregor and Sergeant Bernard Brunnecker are the two other Jasta 29 claimants, both their victims coming down (it is recorded) inside German lines, south of Courrières and at Carvin. Jasta 30 was in this fight too, and lost a pilot over Pont Maudit, near Lens to an SE5a – more than likely Landis's victim.

This same day, Vizefeldwebel Wilhelm Neuenhofen of Jasta 27 had a lucky escape during a fight with SE5s of 32 Squadron. He had four victories to date but would achieve 15. Shortly after 18.00, a fight developed near Monte Notre Dame, down on the French Front, in which Neuenhofen's Fokker was hit in the petrol tank. Leaking fuel he dived earthwards and managed to make a safe forced landing, but he'd been fortunate.

At least two 32 Squadron pilots claimed victories, Lieutenant J N Trusler (C1915) and Second Lieutenant J O Donaldson (E1361), an American who scored his first victory but would go on to score seven. No.32 had come down on the Fokkers which had been attacking some Camels and DH9s, Trusler sending one down on its back. Donaldson claimed his burst into flames and crashed, confirmed by two other pilots, which doesn't look like Neuenhofen, so probably it was Trusler. Of course, Donaldson may merely have seen evaporating fuel and assumed a 'flamer'.

Harold Mellings

One problem in staying so long in combat is that while experience grows, the odds against survival shorten. It is often a balance between this experience, overconfidence and tiredness. Harold T Mellings was two weeks short of his 21st birthday on 22 July but he had been in the RNAS since he was 17 and between 1916 and 1917 had seen action as a pilot with 2 Naval Wing in the eastern Mediterranean, scoring four victories in that theatre of operations. He had received the DSC and Greek War Medal and could have returned home quite satisfied with his efforts. However, he was keen to get back to the war, so after a brief rest, he was posted to 10 Naval Squadron, which then became 210 Squadron RAF. Despite a slight wound from ground fire in April, he had nevertheless increased his score to 15, the last two on the morning of his last day on earth. Already he had received a Bar to his DSC and the new DFC.

Mellings led an evening patrol at 16.15 towards Roulers and ran into Jasta 56 some five miles inland from Ostend. In the mix-up which followed, Mellings (F5914) was last seen down to 200 feet fighting about 12 enemy aircraft at 17.20, and he failed to get home. Second Lieutenant E H Bullen, an American in the USAS attached to 210 (D9626), also failed to get back.

Both men were claimed by Leutnant Ludwig 'Lutz' Beckmann of Jasta 56, one downed at Coolscamp, the other at Kokelere, at 18.20 and 18.22 German time. These were his third and fourth victories; he would score eight in all. Bullen was taken prisoner, while the young Shropshire lad who like so many others had packed so much into his short life, was buried by his enemies at Nieuport.

There was another loss of an experienced flight commander on the northern part of the front on 24 July. Captain P S Burge MC MM was with 64 Squadron, flying SE5s. He had won his Military Medal with the army. Joining 64 Squadron in October 1917, when they still flew the ill-starred DH5, he did not score a victory until March 1918, but then accounted for 11 hostiles, the last on 22 July.

On this evening, Philip Burge led out a patrol at 17.00 along the 1st Army front, flying D6900. He led a lower formation, while Lieutenant D Lloyd-Evans led a higher group. During a general engagement which lasted some 20 minutes, Lloyd-Evans (a future ace, DFC and Bar) was attacked successively by seven and then three Fokker biplanes from above. Meanwhile, Burge dived with his Flight on three other Germans who were joined by three more. Burge was last seen diving at about 10,000 feet firing at very close range on the tail of a Fokker. Other enemy machines engaged the SEs, their number in all being about 15, all working in small units at differing heights. The fight gradually worked east and ended south of Lille, during which time the SEs became scattered.

It seems that the main fight was with Jasta 52, and Burge was shot down by Unteroffizier Murat Schumm, who downed an SE5 south of Hulluch at 19.20(G). He would achieve four victories, this was his third, and survive the war. Hulluch is some way west of the fight zone but obviously Burge was trying to get back when he came down short of the lines.

Leader of Jasta 56 downs an ace
Jasta 56's Staffelführer in July 1918 was Leutnant Dieter Collin who had joined Jasta 2 shortly after Oswald Boelcke's death, in late 1916, so he had a wealth of experience. Earlier he had been with Jasta 22 briefly but after scoring two kills with Jasta 2 (Boelcke), he returned to Jasta 22, bringing his score to six by the beginning of 1918. He was then given command of Jasta 56.

An early warning of British aircraft over the front sent Collin and his pilots off and they soon spotted a number of Sopwith Camels working behind the trenches. These in fact were machines of 209 Squadron. They had taken off at 07.45 on 25 July, in squadron strength (15 aircraft) carrying 20 lb Cooper bombs, which they dropped from 9,000 feet north of Warneton, although they were too high to calculate results. Then they were engaged by about 20 Fokkers which came at them between Houthem and Messines at around 08.45. In the fight which followed, none of the Camel pilots was able to claim any successes but they lost two of their number, Lieutenant J H Siddall (D9636) and Lieutenant A G S Blake (D9621).

Joe Siddall, from St Helens, Lancashire, had been in the RNAS for some time before joining 9 Naval, which then became 209 Squadron RAF. He had nine victories, and was last seen in combat west of Houthem. Collin's claim for a Camel was timed at 09.40 (G), west of Ypres. In fact, Blake was shot down by 'Lutz' Beckmann, his fifth victory.

Raesch bales out
One of the bitterest arguments behind the scenes of the air war was the use of parachutes by aircrew, or to be more accurate, the non-use of parachutes. Parachutes were not new. Almost as soon as men took to the air in balloons, people had been inventing the idea of jumping out of the basket, not just for safety, but for display and amusement. Once the war started, and the balloon and the aeroplane became a feature, the emphasis was on saving life with some form of parachute arrangement. Both sides developed a device for their balloon observers to parachute down in the event of an attack on their floating observation platform, although the apparatus was a bit on the bulky side. In fact the parachute and its pack, was strapped to the outside of the basket, not to the man himself. His act of jumping out and down, pulled the canopy out and, with luck, he floated down to safety.

Because of the size of the apparatus, it was deemed too large and cumbersome to fit anywhere convenient in an aeroplane, but the Germans eventually came up with a smaller sized parachute pack that could indeed be housed with the pilot. These parachutes began to be made available to the Germans in the summer of 1918 and a number of airmen saved their lives with them.

Unfortunately for the Allied airmen, their own leaders did not find favour with using parachutes. Or at least the non-combatant, deskbound bureaucrats didn't. They even suggested that if equipped with a parachute an airman would more likely use it to get out of trouble rather than stay and fight. Boy, were they out of touch! It was that thinking and that attitude that condemned many airmen to death, either in a burning aeroplane or falling several thousand feet in a machine whose controls and/or wings had been smashed. One has to wonder sometimes if war crimes should apply only to atrocities committed on one's enemies.

Perhaps one of the first German aces to save himself with the use of a parachute was Leutnant Josef Raesch of Jasta 43 on this same 25 July 1918. From the southern Rhineland town of Zewen, 21-year-old Raesch had been with the infantry between 1914 and 1917. Training as a pilot he had joined Jasta 43 in June 1918 and scored three kills before this date. On the evening of the 25th, 40 Squadron were patrolling and carrying bombs south-west of Lille. East of La Bassée, at 17,000 feet, they spotted five Fokker biplanes just below the clouds. Lieutenant I F Hind, a 26-year-old South African from Johannesburg, selected the nearest and dived at it, closing to 50 yards before firing a burst of 50 rounds. The Fokker turned over to the left, dived steeply and burst into flames; then the right-hand wings pulled off.

What was not reported, and may not even have been seen, was the pilot kicking himself clear of the shattered Fokker, falling free and then having his headlong fall arrested as a parachute canopy opened above him. Thanking his lucky stars or guardian angel, Raesch landed safely, although he had suffered some burns to his face and hands. It was Ivan Hind's seventh of eight victories. For Raesch, it meant some weeks in hospital, but he was back in action by September. He went on to score four more victories, including the aircraft of two aces, of which more later.

An interesting sequel to this episode, concerns Leutnant Robert Schmitt of Jasta 43. On 12 August 1918, using the repaired parachute which had saved Raesch's life, he too baled out of a burning Fokker and came down safely, except for some facial blistering. How many parachutes have saved two lives? Schmitt too had been shot down by a South African, and from the Squadron of whose compliment of pilots Raesch would kill two aces.

Walter Avery lands a big one

As we have read already, experience and a high score was no guarantee of immunity from the novice who managed to be in the right spot at the right moment. Oberleutnant Carl Menckhoff knew this only too well on 25 July. By this time, Menckhoff, a Westphalian, was no youngster at 35 years old, and had scored 39 victories, gaining for himself the Pour le Mérite. Indeed, his Jasta's score of victories was nearing 60 for no loss. However, on this day he was in combat with the Nieuport 28s of the 95th Aero and was shot down in combat near Château Thierry and taken prisoner. When he asked to meet the man who had conquered him he could scarcely believe what he was told. Second Lieutenant Walter Avery had been on the Squadron for just two days and this was his first combat! Avery received the DSC, Menckhoff got a prison camp.

Mick Mannock's loss

What has emerged over the intervening years since World War One is that Major Edward 'Mick' Mannock was one of the best patrol leaders of that war, and although he did not score the number of victories that some post-WWI writers attributed to him, a list of around 61 was still pretty high.*

Mannock's story is well known. He had been with 40 Squadron in 1917, commanded a Flight in 74 Squadron in early 1918 and was then given command of 85 Squadron after Bishop returned to Canada. He was a very aggressive fighter pilot, with no love for the Germans. Two-thirds of his 60-odd victories were of the destroyed or captured variety. It might also be mentioned that the stories of him and others of giving victories away to some novice pilot to help boost their morale is, for the most part, total nonsense. With a system already in place whereby claims by Allied pilots could be shared, there was no need to 'give victories away'; the novice pilot could be given a share, while the main claimant would still have a shared victory added to his score.

On 26 July, Mannock took out Lieutenant D C Inglis DCM, a New Zealander fairly new to the Squadron, to gain experience. Over the lines Mannock saw and attacked a two-seater of FA(A)292, which he and Inglis shot down. The German crew, flying a DFW CV (2216/18), had taken off at 05.45 for an artillery shoot in the Croix Marnuse area, and were attacked at 06.39 (Inglis's claim was timed at 05.30). There are several

---

* The score of 73 was given by his friend and biographer J I T Jones DSO MC DFC MM, himself a successful air fighter with Mannock in 74 Squadron. It is understood that in writing the biography he was determined to record that Mannock had achieved more victories than the very dubious score of 72 claimed by the Canadian W A Bishop. However, a score around 70 had been noted before, but unofficially.

stories about what happened next, but Inglis (E1294) later made out this report:

> 'While following Major Mannock in search of two-seater EAs, we observed an EA two-seater coming towards the lines and turned away to gain height, and dived to get east of EA. EA saw us just too soon and turned east. Major Mannock turned and got in a good burst when he pulled away. I got in a good burst at very close range after which EA went into a slow left-hand spiral with flames coming out of his right side. I watched him go straight into the ground and go up in a large cloud of smoke and flame.
>
> I then turned and followed Major Mannock back at about 200 feet. About halfway back I saw a flame come out of the right-hand side of his machine after which he apparently went down out of control. I went into a spiral down to 50 feet and saw machine go straight into the ground and burn. I saw no one leave the machine and then started for the line, climbing slightly; at about 150 feet there was a bang and I was smothered in petrol, my engine cut out so I switched off and made a landing five yards behind our front line.'

Inglis force-landed five yards from a British front line outpost at St Loris, having had several hits from ground machine-gun fire through his petrol tank. The machine could not be salvaged and was struck off charge.

Mannock came down between Colonne and Lestrem, his SE5 – E1295 – being burnt out. The machine's engine was a Viper, WD.33703.2603, his guns numbered 19911 (Lewis) and D6522 (Vickers). It was not the most sensible thing to do to fly back across the lines at such a low level and one has to wonder why such an experienced pilot as Mannock should have done so. He actually came down at 07.00 near Pacaut Wäldchen in the area of the German 100th Infantry Regiment, so one must assume it was ground fire from this Regiment that hit the SE5. Despite this, one has to wonder too if Mannock himself had been hit, for being so low he must surely have had time to force-land and get clear before his SE5 began to burn.

Bill Stephenson, the Intrepid

Captain W S Stephenson MC DFC later became well known for his WWII counter-intelligence work in North America, and the famous book of his exploits – *A Man Called Intrepid*. What isn't so well known is his exploits as a WWI fighter ace.

Stephenson came from Winnipeg and had served with the Canadian Engineers, where he was badly gassed prior to joining the RFC. Once a pilot he flew Camels with 73 Squadron, being decorated twice and taking command of a Flight. He gained 12 victories, the last on the afternoon of his final day of combat, 28 July, but the evening patrol ended badly. The Squadron, like some others, had moved down to the French Front, taking part in the attack between Château Thierry and Reims, and subsequently on the whole of the Soissons-Château Thierry–Reims salient. This was nothing new for the unit: it had previously operated alongside the French in June on the Noyon–Montdidier Front.

Stephenson led an evening OP off at 18.20, flying C8296, and got into a fight with German fighters thirty minutes later. In fact they had got tangled up with JGI's Jasta 10, and Stephenson had been hit by Leutnant Justus Grassmann, who downed a Camel. Stephenson was last seen scrapping with seven enemy aircraft over Fère-en-Tardenois and Grassmann's victim came down on the German side of the lines east of Saponay, which is just a couple of kilometres from Fère. Stephenson was not hurt and was taken prisoner. He later made a successful escape and returned to the Allied side of the lines but he did not see further combat. Before his WWII story he had become a successful businessman. He was later knighted, while the Americans gave him the US Presidential Medal of Merit.

Sous Lieutenant Antoine Pierre Jean Cordonnier, a five victory ace with Spa15, was shot down on 28 July. Not knowing the time or location, two possibilities exist: Leutnant Erich Löwenhardt of Jasta 10, his 46th victory, at Fère-en-Tardenois at 15.10, or Leutnant Grimm of Jasta 13, a Spad at 18.50 over Chardun, near Dormans, his first of three kills.

Part of the private war

As has been mentioned before, the RNAS and now the (former RNAS) RAF squadrons up on the North Sea coast, had their own private war with the Navy and Marine pilots. Actions along the coast, around Ostend, Bruges and Zeebrugge, or as far inland as Roulers, saw the DH4s and Camels mixing it with the Marine fighters that were based between Ostend and Blankenberghe.

On 30 July, 213 Squadron's Camels were on a High Offensive Patrol to Bruges, taking off at 11.50, while another, led by Captain J W Pinder who had over ten victories already, followed later. At 16,000 feet the Camel pilots saw five silver-sided Albatros Scouts and attacked. Lieutenant M L Cooper shot at one from point-blank range and saw it go down in a spin, then dive vertically. The Camels then found themselves down to 12,000 feet over Bruges and began to regain height, climbing towards the lines, although three Albatros Scouts followed. About four miles south-east of Ostend, the Camels were about to turn and re-engage when Pinder and his Camels came diving down from the north. John Pinder (D8216) latched on to one Albatros, firing from 50 down to 25 yards, his fire going right into the machine. The Scout commenced a turn in a spiral, still under fire, then began to alternate between spins and dives. Pinder then climbed, joining Lieutenant W E Gray (C76), who was attacking an Albatros head-on. Gray opened fire and the German went into a spin, levelled out at 6,000 feet and headed towards Ostend in a dive.

Either Cooper's or Pinder's opponent was Reinhold Poss of the Seefrontstaffel who managed to get his damaged fighter down for a forced landing. Leutnant Poss had gained five or six victories and would go on to score 11, despite 213 Squadron's attempts at stopping him this July morning. Pinder ended the war with 17 victories; Cooper got six before being brought down by ground fire on 2 October.

John Cowell DCM, MM and Bar
A native of County Limerick, Eire, Sergeant J J Cowell had risen from 2nd Class Air Mechanic observer with 20 Squadron in 1917, then took pilot instruction. With 15 victories as an observer, he had a vast amount of experience of air fighting, and he had been decorated three times. However, upon his return to 20 Squadron as a Sergeant Pilot, he only gained one more victory before being shot down.

The day after his 16th, 30 July, Cowell and Corporal C W Hill, from Birmingham, flew on the evening OP, leaving the airfield at 18.25. The Flight was led by veteran Captain H P Lale. There were 15 Bristols in the formation and they were attacked by about ten fighters north of Armentières but were successfully driven off south-west. Meanwhile, another formation of about 15 scouts came down on the rear of the 20 Squadron group, and as a third bunch joined in, a real dog fight ensued. Lieutenant J Purcell/Pbr J Hills (D7897) saw a fighter on the tail of a Bristol and dived to engage; it went down in a slow spin, but they saw the Bristol going down too and then burst into flames. It was E2471, crewed by Cowell and Charles Hill, from Aston, Birmingham. Other crews claimed one fighter in flames and another out of control.

The main antagonists had been Jasta 16b, led by its commander, Oberleutnant Fritz von Röth, who thus had gained his 18th victory although he had had to spin down and away from Purcell and his observer. Jasta 16 had one pilot wounded in this fight, which took place south of Ypres. Cowell and Hill were both killed, a sad end for such a gallant Irishman.

Also on this 30th day of July, Leutnant Heinrich Drekmann of Jasta 4 was shot down, in flames, at 18.35 over Grand Rozoy. Two GC14 pilots claimed a German fighter near Soissons around this time, Lieutenant de la Poeze, a member of the Groupe's Staff, and Spa75's Sous Lieutenant Bamberger. Drekmann had scored 11 victories and was 22 years old. His last victory had been scored exactly 24 hours earlier, in roughly the same spot.

'Mac' McElroy

It was odd that both Mannock and Captain G E H McElroy MC and two Bars, DFC and two Bars, another Irish veteran, should fall to ground fire. McElroy had flown with 40 and 24 Squadron, scoring 46 victories, lost soon after claiming another, which of course was never credited to him.

McElroy came from Donnybrook near Dublin and was 25. He had fought with the Royal Irish Regiment and been gassed, which caused him to be placed on garrison duty. Wanting a return to action, he transferred to the RFC, became a pilot and then a noted air fighter. Another strange thing, is that both Mannock and McElroy, who knew each other well, had warned each other of being over-zealous in following kills down, or at least chasing them down to a lower level. It had been friendly banter between the two men, but in view of Mannock's loss just five days earlier, the events of 31 July were prophetic.

McElroy had taken off alone on the morning of the 31st in his SE5 E1310, a virtually new machine with just 11 hours' flying time in its log book. It was hard for the Squadron to comprehend when he simply failed to fly home. Some time later the Germans dropped a note to say that he had been shot down by AA fire after shooting down a two-seater – probably a Hannover CL of Schlasta 19 lost this day. He came down and was buried at Laventie. However, Unteroffizier Gullmann of Jasta 56, claimed to have shot down an SE5 south-west of Armentières at 10.15, and Laventie is south-west of this town. It was the German's first and only victory and it seems inconceivable that someone of McElroy's ability should be brought down by such a novice, although it must be said that Gullmann had been with the unit since early February, had survived thus far, and continued to survive until October. Also, this appears to be the only SE5 lost this day. One might be uncharitable and suggest Gullmann saw McElroy fall and made a claim, for it seems equally inconceivable that the Germans would go to the trouble of dropping a note saying how McElroy had died if he had truly been shot down in combat. Wouldn't they too make some mileage out of bringing down someone so distinguished? Another WWI mystery.

CHAPTER TWELVE

August 1918

The Amiens Offensive – the beginning of the end
The German's final attacks on the Marne in July had again ended in failure. The French Front had held. Now it was the turn of the Allies to start the counter-attack, which they did on Thursday 8 August 1918. As can be imagined, the air was full of aircraft this day as the Amiens Offensive began, with casualties to both sides, not only in air combat but by British aircraft brought down during ground attack sorties. Several aces were involved and became casualties.

Captain Roy Manzer DFC was the first. A Canadian from Alberta, although he had been born in Saskatchewan, he had 12 victories flying with 84 Squadron. Flying C8732 on an OP along the Amiens–St Quentin Road at 11.00, he was hit by ground fire and forced to land behind the lines, spending the rest of the war as a prisoner.

Not long after lunch, 54 Squadron were in a fight with enemy aircraft. Lieutenant M G Burger (E5175) attacked a green-tailed Pfalz Scout which he hit, the German going down to make a forced landing on an airfield at Athies. Burger dived down and strafed the fighter, not seeing anyone emerge from it. However, in the same action Captain M E Gonne MC (D6575) failed to return, last seen over Brie at 2,000 feet, at 13.45. It was later reported that Mike Gonne had been brought down and had died of wounds. He had five victories, but whether he was hit by ground fire or an aircraft is unclear.

Lieutenant C M G Farrell, a Canadian with 24 Squadron was a six-victory ace (he would score seven altogether) and was shot down in combat at Warfusée at 15.05, in SE5 D6937. Fortunately he came down on the British side of the advancing troops and his machine was salvaged. He was on his fourth ground attack sortie of the day, and he later received the DFC for this day's work. There were several claims for SE5 machines this day, but none that ties up well with Farrell's combat.

It is the same story for Captain E G Brookes of 65 Squadron. He was last seen near Harbonnières at 14.30 in Camel D8119 and was later reported to have been killed. Eric Brookes had claimed his fifth and sixth victories this very day. The next morning Lieutenant G D Tod of 65 Squadron failed to get back (D1810). An American, George Tod had five victories, but he was later found to have crashed into British lines after being wounded by ground fire.

Paul Billik captured

One of the high scoring German pilots not to be awarded the Ordre Pour le Mérite was Leutnant Paul Billik, commander of Jasta 52, despite his 31 kills, a list started in April 1917 when he was with Jasta 12. He was in fact nominated for the coveted Blue Max but on 10 August he was shot down.

Whether it was thought he was dead, and the nomination withdrawn before he was reported a prisoner, is not known, but it seems sad that such an air fighter should not have the nomination confirmed once it was known he was alive. The mystery too, is who shot him down. He had been in the thick of the fighting over the two days of the offensive, claiming an SE5 and a DH4 on the 8th, another DH4 on the 9th and an SE5 over Herville at 12.40 on the 10th, probably in a fight with 32 Squadron. Billik is reported to have come down inside Allied lines at Vimy–Combles. No.32 Squadron certainly had a fight near Peronne at 11.30 and Combles is just to the north-west. They lost two pilots and a third shot-up. Three out of control victories were claimed, one by Captain H L W Flynn (C1124), one by Lieutenant A A Callender (C9565) and another by the American, Lieutenant J O Donaldson (E5939). It was not only Donaldson's fifth victory but Callender's fifth as well. Was one of these Billik going down? Perhaps coming down unseen and therefore unclaimed as a certain victory, it was not realised in all the front-line confusion that this high-scoring pilot might be one of 32's victories. Things would have been too hectic for even the best Recording Officer to sort out. Jasta 52 also had another pilot injured in a crash this day; was this due to combat damage? Donaldson we have mentioned before, and Jerry Flynn (Hope Laurice Wilfred) from Waterloo, Ontario, aged 19, was killed on 3 September.

Löwenhardt falls

The highest scoring, living German pilot at this stage of the war was Erich Löwenhardt, leader of Jasta 10, with 53 victories. This 21-year-old native of Breslau had commenced his impressive scoring run back in March 1917 under Manfred von Richthofen's tuition and had been given command of his Staffel on 1 April 1918. At the end of the following month, with his score at 24, he received the Blue Max.

In the recent fighting he had downed three Camels on 8 August and two more on the 9th. At a mid-day patrol over the front on the 10th, he and his men got into a scrap with 56 Squadron's SE5s over Chaulnes and he scored victory number 54. The SE5-equipped 56 Squadron had sent out A and B Flights under two Canadians, Captain W A Boger DFC and Captain W R Irwin, Irwin's B Flight providing top cover. They joined up with some Bristol Fighters of 48 Squadron and attacked a group of 15 Fokker DVIIs over Marchelpot. William Boger (B8429), five victories, and another pilot failed to emerge from the battle that began at 11.30 British time. Löwenhardt's pilots later reported their leader downing an SE5 shortly after 12.20 German time, and Chaulnes and Marchelpot are just four kilometres from each other.

However, there are other claimants for Boger. Rudolf Berthold of JGII and flying with Jasta 15 shot down his 43rd and 44th victories at this time, a DH4 and an SE5 over Ablaincourt, the latter timed at 12.30, Ablaincourt being about three kilometres west of Marchelpot. Josef Veltjens also downed an SE5 at 12.45 at Arvillers. So confused was the fighting at this time that it cannot be said for certain which units were fighting which and the fights may well have overlapped. However, after downing his SE5, Löwenhardt then collided with Leutnant Alfred Wentz of Jasta 11. Löwenhardt baled out but his parachute failed to deploy properly and he fell to his death; Wentz parachuted down safely. He had been with Jasta 11 just two weeks and would survive the war.

Berthold too goes down
It was little wonder Berthold was called the Iron Knight. He had been wounded or injured five times so far in this war, suffering untold pain, but flatly refused to give up. Even now his right arm was refusing to heal properly and it hung almost useless, yet he continued to fly, fight and score victories as leader of JGII.

No sooner had Rudolf Berthold gained his 44th victory this day than he was severely shot up (it is also reported that he collided with a DH4). He found his shattered control stick loose in his hand. His Fokker began to dive earthwards. His only chance was to bale out but although he climbed out on to the lower wing, he realised he could not operate the parachute with just his left hand. Then it was too late. His stricken fighter crunched into a house at Ablaincourt, its engine smashing through the roof and into the ground. Berthold was flung into the garden but survived, but his next-to-useless right arm had been broken. After this he was refused a return to operations. If he had been shot down, any one of the aforementioned combats could have resulted in his fighter being damaged.

Another ace brought down this day was M C McGregor DFC of 85 Squadron. As an afternoon patrol ended, McGregor, a New Zealander, had lingered over the front looking for would-be targets and was found himself by three German scouts. Circling for position his machine caught a burst in the engine. Mac went into a spin, followed down by one of the Germans who continued to fire into his SE, smashing his instrument panel but he got down between the two front lines and scrambled clear. Gunfire soon reduced the aircraft to matchwood. McGregor survived to continue scoring, ending the war with 11 official victories and a Bar to his DFC. However, it is again difficult to see who might have brought him down due to the battle area conditions.

Grassmann claims another ace
Justus Grassmann of Jasta 10 had downed Bill Stephenson on 28 July and on 11 August he gained his fifth victory with another ace. No.19 Squadron flew an afternoon OP east of Albert, flying in three staggered formations. The top cover section was attacked by five Pfalz Scouts, and in the scrap one Pfalz was driven down, its pilot taking to his parachute. Captain G B Irving, a Canadian who was about to receive the DFC, was seen to drive down another at 17.00 (E4432), his 12th victory. Re-forming, the patrol continued. At 17.55 the Dolphins encountered more fighters over Flaucourt, Pfalz and Fokkers, and Captain Finley McQuiston became an ace by shooting down one, while another pilot drove down another. However, a dozen more enemy aircraft appeared on the scene, diving upon the Dolphins. Captain C V Gardner claimed one destroyed but one Dolphin was seen to go down in flames and two others failed to get home; a fourth was shot up and crashed on landing.

Jasta 10 was one of the units involved, Grassmann and Alois Heldmann each claiming a Dolphin, over Eclusier by the Somme and east of Frise, at around 19.00(G). Jasta 37 was the other unit that was in the fight, although they claimed no kills. Grassmann probably got Irving, who was killed. Heldmann's victory was his tenth. There is no claimant for the third Dolphin. Jasta 10 suffered no losses, although Jasta 37 did lose a pilot on this date, time unknown. Georg Meyer of this unit claimed an SE5, but with place and time unknown there is no way of knowing if this was the third Dolphin.

Parachute failure claims Pippart
Leutnant Hans Pippart, leader of Jasta 19 burned a balloon near Vrely at 17.00 hours on the 11th for his 22nd victory. He was thus in line for a nomination for the Blue Max, but unhappily for him he did not get back from this sortie. Some record that his machine was damaged in this engagement, while others say that following the attack he engaged a Bréguet over Noyon and his machine was hit in the encounter. Whatever

the truth, Pippart baled out at around 150 feet and the parachute either had no time to deploy properly or it failed. In any event, the 30-year-old Bavarian from Mannheim fell to his death.

If 10 August had been a black day for the Germans, then 12 August was a black day for British aces. No. 209 Squadron lost two on their first show, a High OP of 12 Camels which commenced at 07.30. An hour later the unit began, for them, an indecisive fight with four Fokkers east of Peronne, but lost one Camel in flames. Another machine was seen to go down on fire but nobody was able to see if it was friend or foe. Three more Fokkers were met and fought at 09.00 north-west of Peronne, this time without result. However, when west of Proyart, seven German fighters, with three more circling above, attacked soon afterwards and then more Germans were seen way above. Wop May (F5944), now an experienced fighter pilot, drove down one Fokker, while R M Foster (C61) went down on some fighters attacking a patrol of SE5s south-west of Proyart. Bob Foster's fire damaged the fighter which went down and landed. As he watched he saw the pilot climb out and leaving his engine ticking over, surrender himself to some British soldiers. This was Offizierstellvertreter Fritz Blumenthal of Jasta 53, who was also claimed by one of the SE5 pilots, Captain F R G McCall MC DFC of 41 Squadron; the German was indeed taken prisoner. Another Jasta 52 pilot was wounded.

The main antagonists in this fight, however, were JGI. Lothar von Richthofen was back from his March injuries and had scored nine more kills. In this action he raised this by two more, to make a round 40, but the next day he was wounded again. This wound put him out of the war. Richthofen had in fact knocked down the two aces, firstly Lieutenant K M Walker (D9657), then Captain J K Summers MC (D9668). Walker had gained his fifth victory the previous day – oddly enough this was probably a Jasta 11 pilot. Summers had won his MC with 3 Squadron earlier in the war when it was on two-seaters. He too had scored the previous day, bringing his score with 209 to eight. The Squadron lost one other pilot, Lieutenant D K Leed (B7471), claimed by Leutnant Erich Just of Jasta 11. (It could be, of course, that Just got Walker and Richthofen Leed. If so, Just had just gained his fifth victory over this ace.) I corresponded with Group Captain John Summers in 1968 when he was living in Rhodesia where he had settled during WWII. He recorded:

> 'Those first four days of the battle were pretty hectic and we suffered quite a few casualties both in men and machines. I had to change my machine four times in those four days and there was not time properly to test the guns on the fourth before I took off early in the morning

of 12 August on my disastrous last flight. Walker my fighting partner was killed, also a new replacement (first time over the lines), and I was shot down and taken prisoner by Lothar von Richthofen. The only survivor of my depleted Flight was my deputy leader, Harker, who managed to get his disabled Camel across into our lines.

I was well treated by the Circus and discovered that the two Fokkers Walker and I had shot down the day before included Erich Löwenhardt, one of their stars. The other was a comparative novice and this was obviously the one I got. I dived straight through the Fokker formation, one broke downwards and I shot him down. Meanwhile, another tried to get on my tail but Walker got him.'

(Löwenhardt, of course, had been shot down two days earlier, but Leutnant Max Festler of Jasta 11 had been killed on the 11th, while Jasta 6 had had one pilot killed and another wounded.)

Hind of 40 Squadron

While 209 Squadron were being decimated, 40 Squadron's SE5s were out, a patrol taking off at 08.25. They ran into Fokkers and lost two machines, one flown by Lieutenant H H Wood (D6193), the other by Captain I F Hind (E3984) who had shot down Josef Raesch on 25 July. The German fighters were from Jastas 4 and 5, led by two high-scoring 'Kanone', Ernst Udet and Otto Könnecke. Udet had now taken on the mantle of the highest living German ace, and this was his 53rd victory, scored over Peronne. Könnecke's kill brought his score to 30. Hind was buried at Peronne, Wood being taken prisoner.

First Booker, then Collin

The RAF lost another legendary air fighter on 13 August as the battle continued on the ground. We read earlier of some of the exploits of C D Booker, who was now commanding 201 Squadron, and still flying. If his last morning on earth was recorded sincerely, then he went down fighting.

Charles Booker, the former 8 Naval pilot, now had 26 victories. On this morning he took off at 10.30 (D9642) together with Second Lieutenant Fowles (C143). Fowles was a new pilot. Indeed, the previous day he had been sent off in C125 to try and memorise the local landmarks. After a 45-minute tour around the general locality he had returned, only to damage the machine's propeller on landing. Today his CO would show him the lines.

West of Rosières, just on 11.00, they were attacked by six Fokker DVIIs. Fowler put his Camel into a spin and succeeded in evading them, although his machine was hit several times, while Booker stayed and fought the German scouts, no doubt covering the novice. Booker did not get back. A contemporary account of his last action describes how there were some 60 Fokkers over the front, and ten (sic) attacked him. It is nothing new to have numbers increased when talking of someone's gallant end, but the official report says seven, with no mention of 60, which would have been enough to equip two Jagdgeschwaders. He is also supposed to have shot down three, all destroyed, to bring his score to 29. No Jasta lost three pilots this day, and we now know that the two Camels had been attacked by Jasta 19, commanded by Leutnant Ulrich Neckel since Pippart's death two days earlier. Neckel had been with Jasta 13 and had 21 victories, so he was a good match for Booker. Leutnant Hilmar Schickler of Jasta 13 was shot down and killed in a Fokker DVII over Roye, so we can assume Booker got one of his opponents. Undoubtedly the 'witness' to his last flight may have seen others spin away – the usual ploy of all fighter pilots to get out of trouble. Neckel's victory is timed at 12.00, which is right on time, and it fell at Le Quesnel, which is west of Rosières. It fell into the British lines and the 21-year-old Kentish lad was buried at Vignacourt British Cemetery. Fowler survived the war.

It is highly likely that a Camel evened the score on the evening of the 13th by downing Dieter Collin, leader of Jasta 56. We have read of Collin before but by this time he had raised his score to 13. This evening he was patrolling with his pilots to the south of Bailleul, and so were the Camels of 204 Squadron. Captain A J B Tonks was part of a ten-man patrol of 204 which had taken off at 18.00 to fly a High OP and which was trolling along at 18,000 feet. Adrian Tonks, from London, had been with 4 Naval/204 Squadron since August 1917, surviving a year of combat with six victories thus far; in fact he had just commenced a bit of a scoring run, doubling his score over the next month.

Leading the pilots over Ypres, Tonks saw enemy aircraft at 17,000 feet, one thousand feet lower than the Camels, away to the south-west, five miles south of Bailleul. Both formations began to climb, the Germans trying to get higher than the others, the Camels trying to stay above the Germans. But the German Fokkers climbed better and were soon some 500 feet above the Camels, and seeing one of the Fokkers begin to dive, Tonks 'hoicked' his machine to the right and, turning over the formation, stalled upwards at the EA and fired 50 rounds under its left wing from 100 yards. The Fokker immediately went into a slow spin, losing height swiftly. Tonks followed and attacked again as he flattened out at 7,000 feet over Armentières, 100 rounds at close range. Lieutenant H W M Cumming (D9628) had also dived, and he too began to open fire on the German. Both men watched as the Fokker's spin increased and it was last seen out of control at below 3,000 feet, both men feeling certain it would crash.

The Camels landed back at base at around 20.00, Tonks claiming a shared victory with Cumming, the former's seventh. Collin had been severely wounded, after crash-landing and was rushed to the nearest hospital, where he died soon afterwards. He was buried in Rumbeke Cemetery. Tonks was rested in October, having received the DFC and Bar. He died in an aeroplane crash in July 1919.

Lone man Hewat
The heyday of the lone man was long gone, but sometimes the Dolphin pilots, whose machines could reach a good height (87 Squadron even experimented with oxygen), would climb and patrol while looking for high-flying Rumpler two-seaters. On the 14th, Lieutenant Richard Hewat, who had been with 19 Squadron in 1917 and was now with 87 Squadron, had brought his score to six, and took off for such a lone patrol (E4434). Probably seeing nothing of interest he edged over the lines and failed to return. He was engaged by Leutnant Hermann Leptien, leader of Jasta 63, looking for his fifth victory. He shot down the Dolphin southwest of Bailleul-sur-Berthoult at 11.05(G), Bill Hewat being killed. Flying alone over France was a dangerous occupation. He has no known grave.

Two other aces were shot down on the 14th, but by whom is not so clear-cut. No.20 Squadron's Lieutenant D E Smith (B6987), from Dundee, was probably the victim of Leutnant Marat Schumm of Jasta 56, who shot down a Bristol between Dadizeele and Wervicq at 19.15, Smith being lost in this area between 18.00 and 18.30 British time. David Smith had six victories; Probationer J Hills was in the rear cockpit. Heinrich Kroll, leader of Jasta 24 and Pour le Mérite holder, was badly wounded in the left shoulder on this day, five days after gaining his 33rd victory. He force-landed at Omencourt and was taken to the hospital of Nesle. He did not return to combat duty.

Bill Claxton becomes a guest of the Germans
One of the great Canadian air fighters was W G Claxton, from Manitoba, and despite his 19 years had scored his 37th victory on 13 August. Known as 'Dozy' because he never seemed perturbed whatever the situation, he had been awarded the DFC and been recommended for the DSO.

On 17 August, a patrol led by his compatriot and fellow Captain and flight commander on 41 Squadron, F R McCall in E4018 'F', they had taken off at 08.30 for an OP to the Rangemarck–Nieppe Forest patrol line, Claxton flying SE5 E5910 lettered 'A'. At around 09.00, six Pfalz DIIIs were seen over Armentières and manoeuvring to engage, joined combat at 09.15 near Deulemont. One German fighter went down almost at once and crashed, bagged by another ace Canadian, W E Shields (E4013 'E'), his 12th victory. Some minutes later Lieutenant Barksdale (E3977 'C'), who had become separated, was driven back to the lines. Re-forming the patrol, McCall then attacked a two-seater but the German dived away east. Some other skirmishes followed, and the SEs also dropped some Cooper bombs at a target near Zonnebeke. Then at 10.00 another two-seater was spotted north-east of Bailleul but again the German crew rapidly flew east.

During this time, the patrol had begun to break up. With Barksdale gone and Lieutenant E J Stephens (C5359 'W') having to abort with engine trouble (he was another up-and-coming ace, from Australia), there was obviously a danger of stragglers being picked off. Indeed, Lieutenant T M Alexander (E4014 'O') was suddenly seen going down over Zonnebeke at 09.45 with a Fokker on his tail and did not get back (he was killed). Claxton had already disappeared after the 09.00 encounter, last seen near Ypres at 10,000 feet; he ended up a prisoner. He had actually been badly wounded in the head and only some skilful surgery by a German doctor saved his life.

Claxton had been attacked and shot down by a patrol of Jasta 20, east of Wervicq at 10.00 (G), the victorious pilot being Leutnant Johannes Gildemeister. This was only his second victory (his first had been in May) and he would not score again until well into September, but would end the war with five. Bringing down Claxton must have been quite a feat. Another Jasta 20 pilot bagged his fifth victory in this fight, which must have been Lieutenant Alexander. His name was Karl Plauth and he would achieve 14 victories, command Jasta 51 later in the year, and survive the war.

Johann Schlimpen
The Germans might have been rank conscious, especially regarding commissioned rank and even more so of regular over reserve commissions, but among the lower echelons a pilot's rank mattered little. As long as he could do the job, he could be of any rank. Therefore, while in the RFC and RAF most pilots held commissions, although there were any number of sergeants, that was generally the lowest rank. In the RNAS all pilots had been flight sub lieutenants and above. The French, of course, had more non-commissioned pilots, similar to the German system.

More than one lowly ranked German pilot achieved ace status, and one of them was Johann Schlimpen who was a Gefreiter (private 1st class); he flew with Jasta 45. He was 23 and came from Cologne, and in the summer of 1918 he scored five kills on the French Front but fell in combat on 20 August. He was killed over Fismes at 13.25, probably by Sous Lieutenant Devilder of Spa83. It is possible he was downed by a two-seater, as Sergent Fouquet and Sous Lieutenant Got of Spa278 claimed a scout on the Aisne Sector on this day.

Dennis Latimer is brought down and Ernie Morrow is wounded
Two successful Brisfit pilots were in the wars on the 22nd, the first being Captain Dennis Latimer of 20 Squadron. Latimer, from Dublin, had had a phenomenal run of combat successes since March 1918. Along with his observers his score had reached 28 by this day, 19 of which had been scored with or by his main observer, Lieutenant T C Noel, from Rutland, whose personal score was around 24! Latimer had received the MC and DFC, while Tom Noel collected a MC and Bar.

During August, Latimer had flown with another successful back-seat man, Sergeant Arthur Newland (DFM and Bar), but on the 22nd Noel was once more behind Latimer. They took off on an OP at 0730 on this morning (flying D7993) and over Westroosbeke ran into Jasta 7. Leutnant Willi Negben, a 27-year-old from Zubenhain, got in a telling burst on the experienced pair, killing Noel and forcing Latimer to crash-land and be taken prisoner. This was only Negben's second victory, timed at 0940. He had been with Jasta 7 since May, gained his first kill in June and it had taken two months for him to make it two. He would gain two more before falling in combat near Nieppe on 22 October 1918.

Captain Ernest Theophilus Morrow, from Toronto, was a pilot with 62 Squadron in 1918 and had scored seven victories along with his various observers. On 22 August, his run of luck ended, probably the BF2b claimed by Vizefeldwebel Kister of Jasta 1 although if correct his claim, timed at 11.00, seems a little late, as 62 had flown out at 0605 hours, the fight timed at 0745. It was Kister's second and final victory and was claimed down near Courcelles, although the German was wounded in the action and was sent off to hospital.

They had been escorting 27 Squadron raiding the railway near Cambrai. Morrow was hit in the leg and began to spin down, while his rear man fought to put out flames. This was fellow Canadian Second Lieutenant L M Thompson, who had shared four of Morrow's victories. They both got down without further injury, but it had been close. Morrow in fact had fainted. Louis Thompson used a pyrene extinguisher on the flames while the two-seater (C895) was in a spin, probably hoping (a) the first would go out, and (b) Morrow would regain consciousness. Having force-landed near Ficheux, which is north of Courcelles, Thompson lifted Morrow out of the machine and got help. He received the DFC, as did Morrow.

If it wasn't for the vastly differing areas of combat, one might be lulled into thinking that Negben got Morrow and Kister, Latimer, the times being a better fit, but there is no doubt that Latimer was in action over Belgium and Morrow down between Arras and Bapaume.

Louis Bennett Jr

Bennett was another American flying with the RAF. His was a brief, bright flame that burned intensely for just one month, flying with 40 Squadron. From West Virginia, he had studied at Yale but then enlisted into the RFC in Canada and eventually arrived on 40 in July 1918, going to G E H McElroy's Flight. Between 15 and 24 August he shot down three aircraft and flamed nine kite balloons. In all he flew just 25 sorties, the last on the 24th. On this day, in SE5 E3947, he took off at 12.15 as part of a patrol and bombing mission to the Haisnes area.

He spotted balloons behind the German lines, attacked and burned one at Provin and one at Hantay. British 'C' Battery, RA, then reported seeing an aircraft shot down in flames, and although they thought it was a Bristol Fighter, it seems certain it was Bennett's SE5. He had in fact been hit by the balloon defences, Machine Gun Detachments 920 and 921. Although still alive when German soldiers reached the wreckage, Bennett had broken both legs and had other injuries. He died soon afterwards. His career, so dramatic and so successful, was over so soon that his Commanding Officer had had little time to consider recommending him for an award.

Deadly ground-fire

German ground fire accounted for three other British aces in mid-August: Captain J P Hales of 203 Squadron (five victories) was killed, his Camel (D9671) being hit near Bray on the 23rd; on the 24th, in addition to Louis Bennett, Lieutenant Edgar Taylor of 79 Squadron was killed, his Dolphin D3727 being brought down during a successful attack on a balloon which gave him his fifth victory but cost him his life; and finally 1st Lieutenant L A Hamilton, an American who likewise was hit after flaming a balloon near Lagnicourt at 14.15. Fatally wounded by ground machine-gun fire from the balloon's defences, his 17th US Camel (D1940), crashed. Lloyd Hamilton's tenth victory was his last. He had previously flown with 3 Squadron RAF and he received both the British DFC and the American DSC.

Jerry Pentland

I, as author, have to admit to Jerry Pentland being a particular favourite of mine, although I never met him. When I did learn he was still alive in Australia it was almost too late to do much about it and in any event his sight was all but gone. He died in 1983.

Alexander Augustus Norman Dudley Pentland was also known as Jerry. He was born in Queensland in August 1894 of an Irish immigrant father, while his mother had Scottish parents. After serving with the Australian Light Horse in Egypt and then Gallipoli he transferred to the RFC in early 1916 and became a pilot on a Corps squadron in France. He and his observer scored a victory over a Fokker Eindekker and then Pentland became a scout pilot. Despite breaking a leg playing rugby, by 1917 he was flying Spads with 19 Squadron, won the MC and scored ten victories. He was always to be found in the thick of the action, and on one occasion actually had a British artillery shell pass through the fuselage of his aeroplane! (There are other tales of Jerry Pentland which should perhaps remain under wraps.) After a rest, he became a flight commander with 87 Squadron under Casey Callaghan (the Mad Major); they shared some adventures in London. Returning to France with this unit in April 1918, his scoring continued and by mid-August he had made it 21 victories and been awarded the DFC. However, this all came to a halt on 25 August.

Pentland took off at 08.30 (D3718) on a two-man anti-two-seater patrol, accompanied by Lieutenant D C Mangan (an American C8109). Over Le Sars they found a DFW two-seater at 09.00 which Pentland shot down. But the two Dolphins had been spotted by a patrol of Jasta 57, commanded by Paul Strähle. Strähle later wrote in his diary:

> 'South-west of Bapaume encountered a few Sopwith Dolphins flown by Americans of an American squadron. The Americans were trying to attack German Artbeitflugzeuger (work planes). My Staffel and I dived down upon the Dolphins to drive the Americans off. Knobel and Hechler each shot down one American Dolphin in the region of Kigny. Jenson and I each followed an aircraft to Achiet Le Petit when the American Dolphin looped over in mid-air and glided down towards Achiet le Petit.'

Just why Strähle should record or think that these were Americans is not known. The Americans never operated with Sopwith Dolphins at any time and the fact that Mangan was an American was unknown to him, for although he was brought down he landed in no-man's-land and scrambled to the safety of the British line. Even Pentland got back, although he was wounded in the leg. The bullet had gone through the fuel tank first, so there was also a danger of fire. And he probably did loop in mid-air, who wouldn't after a hit in the leg and a strong smell of petrol in the air? He got over the lines and force-landed. Otherwise the diary note is accurate. Mangan came down near Albert and Pentland recorded the fight as being south of Guedecourt.

Jasta 57 recorded the action as 10.00 German time, with Vizefeldwebel Knobel stating his Dolphin came down on the other side of the lines east of Martinpuich, Flieger Hechler's between the lines south-west of Bapaume. It was their third and first victories respectively. It seems therefore, that Hechler hit Mangan and Knobel wounded Pentland. Both Dolphin pilots also claimed Fokkers shot down, but Jasta 57 did not record any pilot losses. It was the end of Pentland's war, but not the end of an amazing life.*

JGIII wipe out the 17th Aero
As we will see, late August was to witness several big massacres of Allied fighter units, starting on the 26th. The American Camel pilots of the 17th US Aero (the 17th and 148th Aeros flew under British command) were caught by Loerzer's JGIII on this day, the first day of the Battle of the Scarpe, and lost six pilots on an OP between Peronne and Cambrai that afternoon. The Camel patrol had seen another Camel being attacked and went to its aid, only to be dropped on by Jasta 27. The lone Camel was from the other American unit, the 148th Aero, and was also lost.

Of the six 17th Aero machines, two pilots were killed and four wounded. Two of the pilots were aces, 2nd Lieutenant R M Todd (D6595 'N') and 1st Lieutenant W D Tipton (F5951), both captured. Hermann Frommherz of Jasta 27 claimed three over the Vaulx and Beugny areas to bring his score to 15, one being the 148th Camel. Leutnant Rudolf Klimke made his score 16 with another. Bruno Loerzer, JGIII's leader brought Todd down for his 31st victory, at Beugny. Jasta Boelcke's Bolle and Leutnant Heinz claimed their 31st and first kills also near Beugny and over Bourlon Wood. Bob Todd had four victories at the start of this action, a survivor claiming his fifth for him during this combat. Bill Tipton was also credited with two victories in this battle to make him an ace. Rodney Williams, another downed pilot (F5951) taken prisoner, had four victories. JGIII did not lose any pilots.

As we have read already, it was not always air combat that ended an ace's career. On 27 August, Captain C G Edwards DFC of 209 Squadron, who had seven victories, was on a low patrol shortly after midday, in Camel B6371. This was the second such show on this day, both flown over the battle front around Boiry and Hamblain, where Jig-Saw Wood was a feature, between Arras and Bapaume. Edwards' Camel was seen to receive a direct hit from a shell when flying at 200 feet south-east of the Wood. The 19-year-old from St Albans, Hertfordshire, stood no chance.

---

* Read *Australian Air Ace* by Charles Schaedel, Rigby Ltd, 1979.

Heinrich of MFJI

Another Marine ace to fall was Leutnant zur See Bertram Heinrich, a 22-year-old from Charlottenberg. He had been flying with MFJI since early 1917 and had 12 victories by August 1918, having only recently been commissioned.

The Camels of 210 Squadron were operating over the front late in the afternoon of 31 August. They were flying an offensive sweep around Wijnendaele Wood, west of Thourout. They suddenly became aware of seven Fokker DVIIs following a nearby formation of DH9s as they headed back to the lines. They were obviously preparing to attack them out of the sun as the Camels intervened.

Lieutenant W S Jenkins (E4407), a Canadian from Montreal, who had been a corporal in the Canadian infantry, had joined 210 in May. He dived along with the patrol leader, Captain H A Patey DFC, singled out one Fokker and fired 120 rounds at it from 50 yards. The German machine went down in a series of stalls until it passed out of sight in clouds which prevented a conclusive result. It is almost certain that this was Heinrich, although another pilot was also lost over Thourout this date. It was Jenkins' fourth victory, a number he would treble by the end of the war, to receive the DFC and Bar.

Another ace to fall on this final day of August was Captain A J Brown MC of 23 Squadron. Alfred Brown had scored most of his victories with 24 Squadron between September 1917 and March 1918. Following a rest in England, he became a flight commander in this Dolphin squadron and claimed an eighth victory on 23 August. On this Saturday evening he led a patrol out and at 18.45 east of Peronne, along with Captain N Howarth, he ran into 15 red-nosed Fokkers. Some had red fuselages and others had black striped fuselages – JGI machines, the old Richthofen Circus. Although Howarth and Second Lieutenant E J Taylor both claimed Fokkers 'out of control', Howarth's being seen to crash south-east of Mons-en-Chaussée, the Dolphins received a drubbing. Brown went down badly shot-up but he managed to force land on an advanced landing ground. Another pilot went down in flames, a second crash-landed without injury, while two more pilots returned wounded.

Jasta 11 claimed three victories, all timed at 19.45 (G). Only two were actually recorded as being Dolphins, one by Leutnant Friedrich von Kockeritz (his second victory), the other by Oberleutnant Erich von Wedel, the Staffelführer (victory number ten), who saw his go down on the British side, so this was either Brown or the other lad. Leutnant Friedrich Schulte-Frohlinde claimed a 'Camel' on the German side, so this must have been the flamer as no Camels were lost at this time.

CHAPTER THIRTEEN

September 1918

September was to see the largest number of Allied air casualties in a single month of World War One. Bloody April 1917 had held the record until now, but with the coming offensives at St Mihiel and then on the Meuse–Argonne front, the carnage was terrible. For many of the American Air Service personnel, these actions were their first taste of the war and opposing them were some very experienced German fighter units, notably JGII (Jastas 12, 13, 15 and 19) under the leadership of Oberleutnant Oskar von Boenigk. He had taken command on 31 August following the injuries sustained by Rudolf Berthold earlier. Most of the fighting obviously took place down on the French sectors, but the RAF in the northern half of France, continued to keep up the pressure on the Germans around Cambrai, Arras and Ypres.

J O Donaldson and Theodor Quandt
We have read about John Donaldson of 32 Squadron in a previous chapter. Now this American from North Dakota had seven victories but he did not survive the first day of September, although he was not killed. He took off at 12.45, part of an OP to Cambrai in E5939. He, along with another American flying with 32, Lieutenant E C Klingman (B7890), were last seen in combat over Cambrai at 14.40, at 15,000 feet when they should have been starting to head for home. Theodor Quandt was 21, a regular officer from East Prussia who had served since the war started, first in the infantry on both the eastern and western fronts before joining the air service. With Jasta 36 he had scored steadily since the spring of 1917 and now had ten victories. He had been the Staffel leader since 21 August, a post he would retain till the Armistice.

Jasta 36 caught 32 Squadron over the Pronville–Hamel area just before 16.00 (G). As leader, Quandt would normally have had first crack at the opposition, supported by his pilot. He made good work of his attack, shooting down three of the SE5s. Strangely enough, although Donaldson came down south of Pronville and was captured, Quandt's claim for Klingman, who was also taken prisoner, was not confirmed. The third SE5, flown by Lieutenant A E Sandys-Winsch, managed to get back although the pilot had been wounded.

Denied at least his second kill, Quandt went on to score 15 confirmed kills by the end of the war. Serving in the Luftwaffe in WWII he was shot down and killed by French fighters in June 1940 flying a Messerschmitt 109E. Donaldson managed to effect an escape the next day but was soon recaptured. He got away again on the 9th, this time making it into Holland later in the month. He was killed in a plane crash in Philadelphia in 1930.

Stead of 22 Squadron

Ian Stead, a South African, was a pilot in 22 Squadron, and along with his various observers had secured five victories, the fifth claimed in the fight in which he was brought down. He was flying an OP over the 1st Army Front on 2 September (E2516) with Second Lieutenant W A Cowie in the back seat. They took off at 08.25 and an hour later they were in a fight with Fokker DVIIs east of Arras. Their fuel tank was punctured and Cowie was wounded. Stead got the Brisfit down near the front line, west of the Arras–Cambrai road.

Who got them? Unclear, but Vizefeldwebel Albert Lux of Jasta 27 claimed a BF2b near Vis-en-Artois which is right on that long road and must have been by the front line. The BF2b, a new machine with just seven hours of flight recorded, was salvaged and sent to No.1 ASD. Lux also downed an RE8 this day and a few days later brought his score to eight.

70 Squadron takes a beating

There was a file on this combat in the old Air Ministry but it seems to have disappeared. It would have made interesting reading but all we can do now is to record what is known about it.

The date was 4 September 1918. The Camels of 70 Squadron took off at 07.20 to fly an OP into the Douai area. There were 12 of them but only four came back. It was the biggest single loss suffered by a Camel squadron in the war, probably the biggest of any British squadron. Bruno Loerzer's JGIII were flying over the front, and Jastas 26 and 27 had just been in a scrap with 18 Squadron's DH4s. It was just after 08.00 when the Camels arrived on the scene, but Hermann Frommherz had already despatched one 'Four' for his 20th victory. Jasta 26 and Jasta 27 waded in, together with the Geschwader leader and his adjutant, Theodor Dahlmann. The Camels were being led by Captain J H Forman DFC, a Canadian from Ontario and former RNAS pilot of considerable experience; he had nine victories. Other experienced pilots in this fight were Lieutenants D H S Gilbertson, who claimed his fifth in this battle, and Ken Watson, another from Ontario who had gained his first victory on 31 May in the fight with Walter Böning.

It is impossible to say for certain who got whom in this fight, and in addition there was some overclaiming, which is not surprising. Forman was shot down and taken prisoner (C8239) and Gilbertson was killed (E1472). Ken Watson survived, claiming a DVII in flames (E7173) but six others went down, four killed and two more prisoners. Among those killed was the brother of Leonard 'Titch' Rochford, an ace with 203 Squadron. Bruno Loerzer and his adjutant claimed one and two Camels respectively (Loerzer's 37th, Dahlmann's fifth and sixth). From Jasta 26, Leutnant Otto Fruhner claimed three (victories 22 to 24), Vizefeldwebel Erich Buder two (10 and 11), Vizefeldwebel Fritz Classen one (seventh), Leutnant Ehlers one (his second); Jasta 27's Albert Lux got another, for his seventh victory. Jasta 57 had tried to grab some of the glory but instead of claiming victories they only suffered casualties, probably those claimed by the survivors of 70 Squadron. One pilot was wounded with a bullet in the stomach while two others had their aircraft badly shot about. They said there were 23 Camels!

No sooner had this fight ended than another Canadian ace was brought down and taken prisoner, this time a Bristol Fighter pilot of 62 Squadron. This was Lieutenant W K Swayze from Toronto, who had six victories. He had been part of an escort for DH9s of 107 Squadron heading for Valenciennes and had come down on a German airfield at 10.00 after suffering, he said after the war, an engine problem (D7945). However, Leutnant Martin Dehmisch of Jasta 58 claimed him for his sixth victory, on Emerchicourt airfield at 11.00 German time, engine trouble or not.

Captain Herbert Patey DFC
When we read of the Heinrich and Jenkins fight back on 31 August, Captain Patey had been leading the 210 Squadron patrol. Patey himself had raised his own score to 11 by now but on 5 September his war came to an end. London-born Patey volunteered and had served in Egypt and at Gallipoli with the RN Division, but then, as he was found to be only 16, was sent home and discharged. Once he reached 18½, he was called up, in March 1917, trained as a Naval pilot and flew with 210 Squadron for most of 1918.

On the 5th, 210 sent out a Sweep Patrol and at 17.30 north-east of Roulers ran into five Fokker DVIIs. In the fight which followed, 210 lost two pilots, Patey (B7280) and Lieutenant L Yerex (E4390). Both men were later reported prisoners. They had been shot down by two pilots of Jasta 56, 'Lutz' Beckmann claiming his seventh victory at 18.27 (G) and Unteroffizier Ludwig Jeckert his second (of four) two minutes earlier – both south-west of Roulers. Patey returned from PoW camp on Christmas

Eve 1918, only to die in the great influenza epidemic, in February 1919. He was still only 20.

John Doyle of 60 Squadron
Somerset-born John Edgcombe Doyle had served a year in the infantry prior to becoming a pilot. His first unit in France was the famous 56 Squadron but he didn't shine with them and soon moved to 60 Squadron. With 60 he scored seven victories and became a flight commander.

On 5 September Doyle led an OP out at 10.40 to Marquion and scrapped with nine Fokkers. Late in the afternoon he led another OP to the 3rd Army Front helping to escort DH4s, in E1397, and was last seen over the north of Cambrai at 18.25, flying west. With the DH4s below, the SE5s were attacked by 25 Fokker DVIIs, some diving through to get at the bombers which were over Avesnes. The situation was not helped by a heavy rain storm in which the battle was fought. Lieutenant J W Rayner was in Doyle's Flight and when Doyle gave the signal to engage, he dived on four Fokkers that were attacking four DH4s. He shot down one DVII in flames and possibly a second before losing air pressure and having to break off. He saw Doyle going after the damaged Fokker. Another member of the patrol also saw the pilot of a Fokker take to his parachute.

Doyle failed to get back and his pilots noted that he had helped shoot down two Fokkers, one destroyed being shared with Lieutenant O P Johnson. However, nobody had seen Doyle go down. It seems they were in combat with Jasta 4. Leutnant Egon Koepsch claimed an SE5 over Paillencourt, which is north of Cambrai, at 19.15, his eighth of the nine victories he would achieve. This was his second ace, having downed Ken Junor back on 23 April.

Doyle was severely wounded in the leg, so badly in fact he had to have the limb amputated in a German hospital, but he survived. In the 1930s he wrote several articles about his time with 60. He also received the DFC at the end of the war. Jasta 4 lost a pilot, who baled out, but whose parachute had caught fire and he had fallen to his death. It seems he was a particular friend of Koepsch who was so incensed that he strafed the downed British pilot on the ground, but again Doyle survived.

Another ace to fall this day was Lieutenant H B Good of 92 Squadron, whose SE5 (D372) was shot down near Cambrai at 11.10 during an OP, diving down on enemy aircraft. Herbert Good had five victories but it was a bad day for 92 Squadron. Two other pilots failed to get home while another, badly shot about, force-landed east of Vert Galant. It seems they ran into Jasta 37, Leutnant Meyer and Leutnant Himmer both claiming SE5s around noon, at Havrincourt and north-east of Ribecourt. It was

Meyer's 16th and Himmer's second. They lost one pilot killed over Flesquières. Bruno Loerzer of JGIII got the third one over Inchy at 12.10 (G), thereby scoring his 39th victory.

No.4 AFC Squadron lose four

This 5 September Thursday was starting to look bad for the British; 92 Squadron had lost three and one damaged around noon, and later that afternoon the Australians ran into the 'buzz-saw' that was JGIII.

A combined Sweep by three squadrons, including 4 AFC, had taken off at 17.00, led by Norman Trescowthick, from Melbourne (six victories to date). There were four pilots with him, one being Lieutenant Len Taplin from Adelaide, a former army engineer, who upon becoming a pilot had flown in Palestine. By 1918 he was with 4 AFC on Camels. Heading for the Douai area, the fighters got mixed up and the planned co-operation with the other two units failed. Thus the five Camel pilots suddenly found themselves surrounded by a large force of Fokkers – JGIII. They attacked from two sides at once, but Trescowthick could see there was no future without their planned support and gave the signal to avoid combat and head for home. However, either the others did not see the signal or found themselves too heavily engaged to do anything else but fight, Trescowthick suddenly discovering he was heading west alone.

Only Taplin survived, although wounded and a prisoner. The other three Aussies were all shot down and killed. Jasta 26's Erich Classen claimed one for his eighth victory, and Vizefeldwebel Christel Mesch another for his ninth. In Jasta 27, Frommherz got a third, victory number 21, while Lux got the fourth for his eighth. All were timed at around 19.00 (G) in the Henin Liétard–Cuiney and Marquion–Barelle areas. Trescowthick had reported the combat starting at 18.05. From the times of the claims, either Mesch or Lux got Taplin. Taplin was credited with two victories on this day, one during a morning patrol and one in this fight, bringing his score to 12. He received the DFC, returning to Australia after his release.

Dave Putnam, the (almost) unknown ace

Lieutenant David Endicott Putnam hailed from Jamaica Plains, Massachusetts, a descendant of General Isaac Putnam of Revolutionary fame. He had attended Harvard and had enlisted in the French air service while still under age, knowing the USAS would not accept him. Once in France he qualified as a pilot while still only 18, went into the Lafayette Flying Corps (a formation into which American volunteers could be placed and then assigned to French squadrons. The forerunner had been the Lafayette Squadron, but by 1917 there were too many US volunteers and the Squadron, officially N124, was generally full.) From here he was posted to Spa94, then MS156. In the first half of 1918 he scored four official victories and as many as nine probables (the French were always more strict in their confirmations). Putnam then went to Spa38, adding five more certainties – and some seven unconfirmed – to his score. Now he joined the American forces, being made commander of the US 139th Aero, still flying Spads, and brought his score to at least 13. The French had given him the Medaille Militaire, the Légion d'Honneur, Croix de Guerre with Palmes, while from his own country came the DSC. (He was also recommended for the Medal of Honour but this was never approved.) On 12 September he joined combat with Jasta 15 above Limey at 19.35 (G). He was claimed by Leutnant Georg von Hantelmann as his eighth victory. Buried in the Forest of Sebastopol, near Toul, he was later interred at the Lafayette Memorial, Parc de Garches.

Von Hantelmann downed another ace four days later, the 16th, this time the great Frenchman Maurice Boyau of Spa77, victor in 35 combats, the last on this very day. Maurice Jean-Paul Boyau had lived in Algeria, and was 30 years old. A pre-war infantryman he also captained the French rugby team. He was assigned to the service corps when war came but eventually managed a transfer to aviation. Flying Nieuports and later Spads with his squadron, he had scored steadily since March 1917, became an Officer de la Légion d'Honneur and also received the Medaille Militaire and Croix de Guerre with at least 14 Palmes and one étoile. He went down south-west of Conflans at around 11.25 moments after being seen to down his 35th opponent.

Wessels of MFJI
In our book on German aces, *Above the Lines* (Grub Street, 1993), it was stated that Heinrich Wessels' birthdate was given as 18 April 1877, in which case he was 41 years old in 1918. So far nobody has challenged this date.

He had become a member of MFJI at the start of 1918 and claimed six victories by the early summer. He was shot down on 16 September at Oberschtneweide, and succumbed to his injuries. He had met 204 Squadron along the North Sea coast.

Three formations of Camels totalling 19 machines had flown off at 17.20 and crossed the lines with C Flight below, B Flight above, and A Flight high up as top cover. A balloon was supposed to have been attacked but they could not find it in the haze. At 19.00, having seen little else, the Camel pilots were about to return but then spotted five Fokkers east of Ostend. B Flight went into the attack but the Germans were suddenly joined by more enemy fighters. A strong west wind drove the dog fight over Blankenburghe and Zeebrugge, and it was getting dark, due to poor visibility mainly. In the fight it was estimated there were 14 Fokker DVIIs, four monoplanes (presumably Fokker monoplanes or Brandenburg W29 seaplanes) and three two-seaters. Captain C R R Hickey (F3942) attacked a Fokker which went down in a spin and burst into flames. He then sent another down out of control, but was then attacked by a biplane and a monoplane seaplane. (These were probably the first DVIIIs seen by the British Squadron, unless they were indeed Brandenburg seaplanes.) Freeing himself from these, he then attacked another biplane and this too went down in flames just over the sea.

Lieutenant W B Craig (D3374) fired at a biplane which lost its lower wing. He then dived to help a Camel with a fighter on its tail, and as the biplane overshot Craig fired 200 rounds into it and it fell away in a spin. With the sky full of aircraft, Craig hammered at another biplane and this too went down out of control. He then saw Hickey destroy another in flames. Lieutenant R M Bennett (D8187) had been the pilot saved by Craig and he went into a stall but levelled out lower down and headed south, meeting another Fokker head-on. After a brief tussle, he shot it down into the sea, seen and confirmed by another pilot. Lieutenant F G Bayley (D8146) attacked a salmon-coloured DVII, saw one falling in flames but his own guns jammed as he made an attack. Clearing his guns he attacked a monoplane seaplane on Hickey's tail and drove it off. A few minutes later Bennett spotted an LVG and shot it down too. Meanwhile, Lieutenant P F McCormack (B7254) saw a two-seater on the tail of a Camel, fired 150 rounds at it and saw it spin into the sea off the Zeebrugge Mole. He was then hotly engaged by several Fokkers but managed to spin down and out of trouble.

Apart from Wessels, MFJI had two pilots wounded, one only lightly. Certainly 204 claimed three fighters shot down for no loss, plus a two-seater. Thus it can be assumed that either Hickey, Craig or Bennett brought Wessels down. One of Charles Hickey's kills was shared with Bennett, Bayley, Second Lieutenants N Smith and J R Chisman, but Hickey had raised his personal score to 17 in this fight. William Craig's three claims, one destroyed and two out of control, made him an ace as did Risdon Bennett's claim for a Fokker and the LVG.

Spa154 lose two aces
Jacques Louis Ehrlich of Spa 154 was another balloon specialist. Since June 1918 he had flamed 17 and downed one aircraft. He had already received the Medaille Militaire and had just been nominated to be a Chevalier de la Légion d'Honneur, but on 18 September, in getting his

19th victory, he was brought down. Ironically, the next day came his promotion to commissioned rank.

On the same sortie, Adjutant Paul Armand Petit, six victories including three balloons, was killed. Despite the difference in scores, Petit already had become a Chevalier de la Légion d'Honneur. He fell over Brimont on the Marne front flying Spad XIII No.15060, after assisting Ehrlich in the destruction of the Brimont balloon at 18.05, thereby gaining his seventh victory, but at high cost.

Back in April, three days after gaining his first victory, Petit had been forced to land after a combat with Fritz Pütter of Jasta 68 near Moreuil, but had survived, and on the right side of the lines. He was Pütter's 17th of 25 victories. Pütter met his end in the summer after his (faulty) tracer ammunition ignited in his Fokker on 16 July. Badly burned as a result he died on 10 August.

As far as is known, both Frenchmen were caught by the balloon ground defensive gunfire. Although Petit was killed, Ehrlich managed to get down to be taken prisoner. He lived until 1953 in Paris, the city of his birth. He was 60.

Joe Wehner
The story of Joe Wehner and his friend Frank Luke Jr, pilots with the American 27th Aero Squadron, has all the drama, passion and adventure to make a great TV movie. We shall read of Frank Luke later, although all aviation historians know something of his story.

Wehner, like Luke, was of German descent, born in Boston in September 1895. Coming to the 27th Aero, he teamed up with Luke, both men being loners, outsiders really, but for a few brief days they became what Americans today might call 'hot property'. In those September days, Luke and Wehner made life hell for the German balloon observers, for between them they burned more than a dozen, and shot down a few aircraft. Wehner often played escort and guardian, and after helping Luke to flame two gasbags on the 18th, bringing his own score to five (his actions had already resulted in recommendations for the DSC and Oak Leaf Cluster), Wehner took on a number of Fokkers, helping to protect his friend. The Fokkers were from Jasta 15, Georg von Hantelmann downing the Bostonian at 16.30, the latter's Spad XIII (7555) falling west of Conflans. He was von Hantelmann's 16th victory.

Fruhner bales out
Two conflicting reports as to how Otto Fruhner was forced to bale out on 20 September are open to debate. JGIII got into a series of fights all around Cambrai. The Dolphins of 87 Squadron and the SE5s of 85 too,

while Camels of No.203 joined in. The Dolphins claimed three, and 85 another. 203 recorded two definite claims and one out of control.

The Camels were at 12,000 feet above Haynecourt at 15.30 when they joined in. Lieutenant W H Coghill (D9640), in one of his first actions, dived on a Fokker firing 50 rounds into it at point-blank range. The wings of the machine folded up and the German crashed. Lieutenant H W Skinner (D9658) crashed another, while Lieutenant F J S Britnell (D9611) claimed one out of control. Fruhner was attacking another 203 Squadron Camel, flown by Second Lieutenant M G Cruise (E4409), and in shooting it down is said to have collided with it, or that the doomed pilot had actually rammed his Fokker. Nothing like this was reported by the British pilots, but that is not to say it didn't happen, either as a mere collision, or as a last desperate ramming action by Milton Cruise, a 23-year-old from Port Dover, Ontario, Canada, who was killed.

Fruhner certainly claimed a victory in this fight, his 27th, and Leutnant Franz Brandt downed another for his 10th. Second Lieutenant C G Milne came down and was taken prisoner (E4377). Fruhner baled out of his smashed Fokker and landed safely but he did not see further combat. He was also nominated for the Pour le Mérite but never received it. Leutnant Schneider was also wounded in this action but by whom is not clear. Coghill, from Doneybrock, Dublin, was himself brought down on 26 September during a raid on the airfield at Lieu St Amand in Camel D9640. After the attack, just on noon, he was on his way home when he was engaged by a Fokker DVII and was forced to land north-east of Cambrai, possibly shot down by Leutnant Ernst Bormann of Jasta Boelcke who claimed a Camel north of Cambrai on this date.

Another German ace knocked out of the war this day was Leutnant Hans Böhning of Jasta 79b. He was attacking an unknown DH9 over Soriel in Fokker DVII 747/18, but was wounded in the hip by return fire and had to break off the action. He too did not see further combat, and his score remained at 17, achieved in just over a year.

Klimke is wounded
It was a bad few days for the German aces. First Fruhner, then Böhning and then Rudolf Klimke. Klimke had been in the 17th Aero's fight back in August and his score whilst with Jasta 27 had reached 17 too. Again it is unclear as to the events, but the Jasta was certainly in a scrap with 209 Squadron's Camels at around 18.35 near Ecourt St Quentin. Wop May was leading his Flight (in D9599) and had been out for an hour.

They encountered seven Fokker DVIIs and May attacked one and sent it down but he could not watch it as another was attacking Lieutenant G

Knight (F3233). May came to the rescue and then both men fought the Fokker down to 1,000 feet. Another Fokker then went for May, Knight firing into it whereupon it spun down, burst into flames and crashed between Ecoust and the Sensée Canal. May's Camel had taken a beating and he was lucky to get it back. This may have been the Camel that Leutnant Friedrich Noltenius of Jasta 27 claimed as (apparently) force-landing near Gavrelle at 18.30. This victory was not confirmed. Klimke also made an unsuccessful claim in this action. Jasta 27 had one pilot wounded around this time – Klimke. Was he the one hit by May and Knight? He is certainly not reported to have crashed in flames, and there is a report that he was hit by three bullets in the shoulder during an attack on a Bristol Fighter. No Bristols were shot down this day and one can never know if Klimke had misidentified another British two-seater. However, Klimke was the only recorded casualty this 21 September, and his wound put him out of the war.

Timbertoes Carlin

The Germans pulled one back on the 21st. Captain Sydney Carlin MC DFC DCM was a flight commander in 74 Squadron. He had won his DCM and MC, but he lost a leg while in the Royal Engineers in 1916. Obviously in no shape to run around the trench system, he volunteered for the RFC and by 1918 was with the Tigers of 74.

With ten victories and a DFC he had obviously found a niche for himself when most men would have been happy enough to sit out the war either at home or behind a desk. Carlin, or Timbertoes as his pals called him, was made of sterner stuff. However, this all came to an end on 21 September. The Squadron sent out an OP of five machines to Lille that afternoon, and at 18.40 they met Fokkers. Two were claimed as destroyed, one having its rudder and tail shot off, another with its bottom right-hand wing folded back spun and broke up, while two more were claimed as out of control, one leaving a smoke trail. Carlin (D6091) was the only casualty. His victor was Unteroffizier Siegfried Westphal of Jasta 29, the second of six victories he would have by the war's end. The German claimed him east of La Bassée, near Hantay, at 18.45, so Carlin had nearly got back. Carlin served again in WWII, flew several Ops as a rear gunner in a Wellington, but in May 1941, while a gunner with 151 night fighter Squadron (Defiants), he was killed in an attack on his airfield, aged 48.

Louis Strange's brother Ben

By 1918 Colonel Louis Strange DSO MC was pretty well known among the RAF for he was leading one of the first fighter wings, No.80, comprising Camels, SE5s and Bristols, plus a DH9 bomber squadron. One of the early war fliers, he had been among the first to equip his aeroplane with a machine-gun back in 1915. His younger brother Ben, actually Gilbert John Strange, was a fighter pilot with 40 Squadron. On the morning of 24 September he had six victories, and had just recently been made a flight commander.

Soon after dawn the SE5s took off for an OP and bombing sortie, between Cambrai and Douai, Strange in E4054. Flying between these two towns, near Abancourt, they were engaged from behind by several Fokker DVIIs that came up from below and were actually stalling in order to fire up at the British scouts. An SE pilot attacked one and claimed it shot down leaving smoke from the cockpit area, while others reported Ben Strange shooting one down in flames, but that another DVII got him. He was seen to catch fire and go down, although seemingly under control. If he was still alive it was probably a painful and prolonged death as he headed down. There is generally confusion in any combat so whether these two Fokker claims were one and the same machine is open to debate. What seems clear is that Strange was attacked and set on fire by Leutnant Martin Dehmisch of Jasta 58, credited as his tenth victory, but that he was almost immediately shot down himself. Although Strange was given credit for his seventh victory, more than likely Dehmisch was actually shot down by the other pilot, Second Lieutenant G Stuart Smith (E3046) who would score three victories by the Armistice. Dehmisch crashed and was soon in hospital severely wounded, dying the next day, the 25th. Smith reported:

> 'I observed a Fokker stalling up on the left to secure a favourable position to attack. I did a climbing turn to frustrate this and saw another EA stalling up on the right. I immediately did a half roll to the left and got a full drum of Lewis and 100 rounds of Vickers into the first EA at point-blank range. EA went down out of control, spinning and side-slipping, with smoke issuing from the cockpit. I was unable to follow EA owing to being attacked by two other EA and driven down for about 4,000 feet.'

Bill Craig goes down to MFJV
Lieutenant William Benson Craig of 204 Squadron, who had been in the fight in which Wessels of MFJI had been killed ten days earlier, was himself brought down on the 26th. Captain Charles Hickey DFC had again been leading almost the same group of his pilots, escorting DH4 bombers back from an attack on Bruges just before 11.00, together with a second Flight. Seven Fokkers climbed up under the tails of the Camel

Flight behind Hickey's, but Hickey saw them and turned to attack, followed by the others. He saw one Fokker attacking a Camel so went for it, firing from close range. The Fokker began smoking and went down out of control. In the fight which followed, another DVII was seen to crash into the sea north-east of Blankenburghe while three more were claimed out of control.

The Fokker pilots were from MFJV and they claimed two Camels, one by Flugmaat Christian Kairies (believed to have been Craig), a second (C75) by Flugmaat Karl Engelfried. These were their seventh and fourth victories respectively. Engelfried too would gain five victories; MFJV suffered no pilot losses. The 23-year-old from Ontario, formally with the Canadian Field Artillery prior to transferring to the RFC, had been flying D3374 and had secured his eighth victory two days earlier. His DFC was announced after his death.

Udet wounded

Oberleutnant Ernst Udet had now become the living ace of aces on the German side with 60 victories. The 22-year-old from Frankfurt had had a long war despite difficulty in actually joining the armed forces before the war. When war came, however, he was assigned to a Württemberg army division as a motor cyclist. Taking flying lessons at his own expense he began military flying in 1915 and as a lowly Gefreiter flew two-seaters with an artillery unit. He eventually became a Kek pilot and scored his first aerial success in March 1916 during a raid on Mulhausen by French and RNAS aircraft. Commissioned in early 1917 he served with Jasta 15 and then 37 and as his score mounted he was noticed by von Richthofen, who had him assigned to Jasta 11 in early 1918. He later commanded Jasta 4, having received the Pour le Mérite. Over the summer battles of 1918 his score increased steadily until it reached 60 on 22 August.

These victories were not achieved without some retribution from the Allied fliers. Several times his aircraft was hit and on 29 June 1918, while attacking a Bréguet over Cutry at 07.40, the rear gunner hit his Fokker DVII. His controls were shot away and he was forced to bale out. He fell into the front lines, lucky not to have come down behind the French positions which, had he done so, would have ended his scoring run at 35.

On the afternoon of 26 September he was credited with two further victories to make his war total 62. There seems to be some confusion about the victories claimed and credited on this day, for although Jasta 4 and Jasta 77 both claimed four DH9 bombers, it seems that Kofl 19 gave credit to Jasta 77, although there is a photograph of one Jasta 4 pilot standing beside a downed DH9 of 99 Squadron, the unit they were in action against. In fact, 99 Squadron, of the RAF's Independent Force, lost five of their bombers on their raid upon Metz–Sablon, with two more badly shot-up. Whatever the outcome, Udet was credited with two kills, even if it was a case of bending some rules to add to his impressive record. However, he was wounded in this action, 99 Squadron claiming two driven down and another in flames at around 17.10, but it is obviously unclear if any of the missing crews also shot down aircraft before falling themselves. All that can be safely said is that 99 Squadron wounded him; they also wounded a Jasta 77 pilot, but it was only a slight thigh (buttock) wound. Udet was back in early October, only to be taken off flying on 22 October and after some leave was posted to the Inspektorate of the Flying Corps (Idflieg).

CHAPTER FOURTEEN

September – October 1918

Black September

As mentioned at the beginning of the last chapter, September 1918 was to see the heaviest casualties of the war inflicted on the British, French and American fliers during WWI. More than once in August and September the deadly German units had taken out six, sometimes eight aircraft at a time from a unit. I have only mentioned those actions in which aces were lost, but there were others. American Spads of the 13th Aero had four of their number knocked down on 14 September by Jasta 18. The American 11th Aero lost five DH4s on 18 September, set upon by Jastas 6 and 12. No.213 Squadron lost four pilots on 25 September, three in a fight with MFJII; 55 Squadron, DH4s, lost four and others shot up to pilots of Jasta 10 and 19 on the same day. It seemed as if the Germans could inflict casualties almost at will. On 16 September, overall losses were 31 British, five American and nine French aircraft lost, a total of 45. That the Allied side claimed over 100 German aircraft this same day is laughable, and sad.

With the Battles now raging at St Mihiel and the Argonne, many of the Allied casualties were among the mostly untried Americans, whose courage and bravery were unquestioned, but the German fighter pilots were just too knowledgeable and experienced. The Germans shot down over 550 Allied aircraft in this deadly month. The Allies claimed over 1,000 victories, a possible ratio of 10 to 1 of actual German casualties.* There is no getting away from it, the German fighter pilots were good at their job. And over the final days of September, aces seemed to fall like rain. Friday 27 September saw the attack on the German Hindenburg Line begin, followed the next day by the offensive in Flanders. The air was full of aircraft, bombing, ground strafing, and in combat.

Paul Strähle wounded

The leader of Jasta 57 had a score of 15, but after 27 September did not score again. He led a patrol to intercept bombers (in Fokker DVII 4025/

---

* Read: *Bloody April . . . Black September* by N Franks, R Guest and F Bailey (Grub Street, 1995).

18), climbing between Douai and Cambrai. In his diary he recorded:

> 'Climbed in the direction of Arleux in order to cut off English formation flying towards Cambrai. Leutnant Blum warned of English aircraft above in some clouds – DH4s – heading back to the front. Climbed to engage but then abandoned it and headed for the lines. Joined up with Jasta 28 and another Fokker staffel near Souchy Lestrée and headed for Cambrai. Nothing seen so turned and climbed north. Saw two machines approaching – DH4s. I got close to one and opened fire from underneath. Suddenly I felt a strong hit on my left forehead; I must have become unconscious for a moment for the next moment I was gliding down.'

Another bullet hit the cocking handle of one of his machine-guns as it was moving and probably saved Strähle a worse hit. These DH4s may have been from 27 Squadron, who lost one machine over Sancourt (crew PoWs) to Jasta 57's Leutnant Johannes Jensen, for his fifth victory, downed at Fechain, just to the north of Sancourt.

In fact 27 had a mixed force of DH4s and DH9s and recorded the combat just north of Emerchicourt. Lieutenant F C Crummey and Second Lieutenant F F McKilligan (B2135) reported being attacked by a Fokker, into which McKilligan fired 70 rounds at close range. It was seen to go down apparently out of control, the two men watching it for about 4,000 feet before it fell into some cloud. Emerchicourt is 13 kilometres north of Cambrai.

A British ace lost this same morning was Second Lieutenant P S Manley and his observer, Sergeant G F Hines (C944) of 62 Squadron, on an OP east of Douai. They were last seen going after a German aircraft and failed to get home. Pat Manley had scored his fifth victory on the 24th; in fact all his claims had been during this month. They ended up in a prison camp. There were several claims for BF2b machines this day, but none ties up with time and place, always assuming, of course, they were brought down by a German fighter.

Fritz Rumey is killed
The third ace to go down this morning was the great Fritz Rumey, the 27-year-old high-scoring pilot with Jasta 5, having downed his 43rd and 44th victims the previous day. All earlier indications pointed to his loss being attributed to a collision with the SE5a of Lieutenant G E B Lawson of 32 Squadron, part of an OP to Emerchicourt escorting 27 Squadron's bombers. However, this action was at around 09.30, and Rumey claimed

his 45th and final victory at 12.05, east of Marquion (the British and German times are the same remember). Apart from the time, the two locations are some 16 kilometres apart.

Jasta 5 appear to have been fighting Camels of 54 Squadron east of Marquion, and they lost two, Jasta 5's Rumey and Josef Mai each making a claim. It seems reasonable to assume that Rumey was lost in the afternoon on another sortie. He apparently parachuted from his damaged Fokker over Neuville St Remy, Cambrai, but the parachute failed to open. This too is a long way from Emerchicourt, 14 kilometres.

It is difficult with the passage of time to know if the collision story was known at the time, or later, when people were trying to discover how he was lost. Did someone pick out the fact that Lawson had collided with a Fokker that day and assumed it to be Rumey? It wouldn't be the first time someone had tried to show that a pilot could not have been out flown, merely lost due to bad luck. However, if there was a collision it was certainly not between Rumey and Lawson. The SE5s of 32 were in combat again this same afternoon, and the fight was right above Cambrai. Captain F L Hale, an American, claimed three Fokkers at 17.20, one destroyed and two out of control. Is this where the 32 Squadron story began perhaps?

Captain C V Gardner, 19 Squadron
Cecil Gardner was a flight commander flying Dolphins with 19 Squadron and had just recently received the DFC. On this 27 September he shot down his tenth enemy machine in the morning but on a patrol in the late morning, he failed to extricate himself from a fight with Fokkers.

The OP had begun at 11.20, providing cover for DH4s. At 12.50 ten German fighters dived on the bombers and the Dolphins attempted to fend them off over Aubigny, near Bapaume. Gardner (E4501) got his damaged machine down, having been badly wounded. Although his Dolphin survived and was sent to 1 ASD for major repair, the 29-year-old Gardner, from Buckinghamshire, died of his wounds three days later. He was brought down by Unteroffizier Gustav Borm of Jasta 1, timed at 13.00 in the fight which had started over Bourlon Wood. It was his third victory and he would end the war with five.

'Siffy' Thompson goes west
Captain Samuel Frederick Henry Thompson MC DFC of 22 Squadron downed victory number 30 on the morning of 27 September, making him the highest scoring Bristol Fighter pilot after the Canadian, Andrew McKeever. And at least 18 of these had been scored with his front Vickers gun. His usual observer was Sergeant R M Fletcher DFM who had been behind him while they had notched up 25 victories.

On this fateful day, Thompson had someone else in the back, but it was no novice observer; Second Lieutenant Cliff Tolman had eight victories of his own. No.22 had launched a patrol over the 1st Army front at 14.30 and Thompson (E2477) had last been seen over Cambrai, still flying east. The Bristols got into a running fight with a mixed bunch of German fighters from three different Jastas, 5, 40 and 46. Each unit claimed one two-seater: Jasta 40's Carl Degelow brought one down west of Valenciennes, for victory number 18; Oberleutnant Otto Schmidt got another down at Catternières, south-east of Cambrai, at 15.50, his 16th; then Vizefeldwebel Oskar Hennrich bagged another over Neuvilly at 16.00, his 18th.

Some records indicate Degelow claimed a machine of 88 Squadron, but he reported its crew were taken prisoner, whereas the 88 pair were killed. Only one 22 Squadron crew were captured, another had its pilots killed and observer wounded, while Thompson and Tolman both died, therefore either Hennrich or Schmidt must have got Thompson, who was 28 years old and came from south-east London. The Brisfit crew have no known graves.

The last two aces in trouble this day were the German commander of Jasta 59, Oberleutnant Hans-Helmut von Boddien, five victories, and Max Näther, CO of Jasta 62. Von Boddien had flown with Richthofen's Jasta 11 in 1917, but then became leader of Jasta 59 despite having scored not a single victory. His five with Jasta 59 came between March and September but he was severely wounded on the 27th with a bullet in the calf. He had been attacking a DH9 and was hit by the observer. This may have been in a fight with DH4s and 9s of 98 Squadron early in the day – one of his pilots claimed a 98 Squadron bomber – or the fight with 27 Squadron a little later in which 32 Squadron had featured. If the latter, then Second Lieutenants C M Allan and C E Robinson (C1212, DH9) claimed a Fokker over Emerchicourt, although they said their victim was going down on fire.

Max Näther, just turned 19, bagged his 17th victory on this day, a Spad XIII, in a fight with Spa77 over Montzéville, but another Frenchman had put a burst into his machine and he'd been slightly wounded. He flew home all right and remained with his unit, so in all probability the French Spad pilot had not even made a claim.

Ernest Hoy

Vancouver-born Ernest Charles Hoy, 23, was a flight commander with 29 Squadron in 1918, and in just over a month had claimed a dozen victories and had won the DFC. On 28 September he was leading a formation of

SE5s into the Menin area towards the late afternoon and ran into three batches of Fokker DVIIs. They had left base at 15.40 and Hoy was last seen south-east of Menin, in combat, at around 16.40 in C1914. The fight had been with Jasta 43, and although the British pilots claimed two Fokkers crashed and another in flames, there appear to be no pilot losses in this area. Hoy, however, was lost, brought down by Leutnant Josef Raesch at 17.50 east of Ypres. Raesch, it will be recalled, had baled out of his burning Fokker back in July, but now he had achieved his fourth victory. No.29 Squadron would meet Raesch again in October.

Another bad day
Sunday 29 September saw several more aces in trouble. On a dawn show, Lieutenant Ross Macdonald of 87 Squadron took off as part of an OP in C4155 'Q' and didn't get back. He had flown this Dolphin to France in April and scored all five of his victories in it. From Winnipeg, he was a little older than some, 29, but he had already a long record behind him, having been an observer in 15 Squadron 1916–17, before training as a pilot. He went down in a fight with several Fokkers near Estourmel but there is no record of anyone claiming a Dolphin, so it is difficult to see who might have got him, always assuming he was brought down and did not have engine trouble, or was hit by ground fire afterwards. He was last seen south-west of Cambrai and ended up a prisoner of war.

Jagdgruppe leader Gandert goes into the bag
Oberleutnant Hans-Eberhardt Gandert, as leader of Jasta 51, also commanded Jagdgruppe 6. This former Jäger Battalion soldier had been commissioned in 1912 when he was coming up to 20. Now, six years later, he had celebrated his 26th birthday earlier this month. First in two-seaters, he'd been shot down on the Russian Front in late 1914 but got back after three days. He had continued flying in Russia and later over Romania but at the end of 1917 took command of Jasta 51 on the Western Front. Two kills in the East had now increased to eight, including 'Lobo' Benbow in May as we read earlier. On this Sunday, Gandert had attempted to attack the balloon lines at Langemarck, and although the Jasta records note an unclaimed balloon victory for him on this day, he was brought down in that area and taken into captivity.

This morning also saw two Bristol aces go down. The first was Lieutenant N S Boulton (with Second Lieutenant C H Case) of 20 Squadron, who was last seen over the lines heading west. In E2561 they had taken off at 09.05 as part of an OP. It had been thought they had been brought down by Josef Mai of Jasta 5, but it now appears this was an 11 Squadron machine, as the time for Mai's kill does not tally with Boulton's loss. They were probably the 20th victory of Friedrich Altemeier, Jasta 24s, who downed an aircraft at Montbrehain at 10.40, that is near the area in which Boulton came down, being buried beside the wreckage of his machine on 3 October. Charles Case also died.

The other Bristol (E2517) was flown by another 22 Squadron pilot named Thompson, this time Lieutenant C W M Thompson, 12 victories. He and Lieutenant L R James were both taken prisoner, last seen in combat east of Cambrai. They had taken off at 10.00 and engaged Fokker DVIIs near Cambrai, probably shot down by either Unteroffizier Karl Fervers or Vizefeldwebel Paul Keusen of Jasta Boelcke, victories three and one respectively.

Jasta Boelcke's Paul Bäumer was also scoring well this day, gaining three kills, a BF2b, a Camel and an SE5, to bring his score to 38. The Camel was flown by Captain E B Drake, a five-victory ace flying with 209 Squadron. He too had been flying on the day Manfred von Richthofen was shot down, having been with the unit when it was 9 Naval. He had also shot down a Gotha bomber while in England in August 1917. In addition he had also shot down a Jasta 46 pilot on 27 June from which the man had made the first successful parachute descent in combat.

The Squadron had sent out a low patrol of 12 Camels to the south of the River Scarpe area at 12.30 and Drake had not been seen since 13.15 over Cambrai, Drake leading his Flight in E4376. They had crossed the lines at 6,000 feet and dived down through clouds, finding several trains in a station to the east of Cambrai. All aircraft had attacked with bombs and machine-gun fire but poor visibility at low level made it difficult to assess results. Cambrai itself was enveloped in smoke. One Camel failed to return and it appears that Bäumer had spotted the straggling Sopwith and attacked. It had gone down south of Sailly, just to the west of Cambrai, his 38th victory, although it may still prove to have gone down by ground fire from the attack on the station. Ed Drake has no known grave.

Frank Luke's last fight
We read earlier of Frank Luke and Joe Wehner. Frank had lasted a few more days, even more personally withdrawn and deeply saddened by the death of his friend. Nothing or nobody seemed to be able to stop him taking off to engage the enemy, and his chief target was always the German balloon lines.

On this late afternoon, he flamed three balloons at Avocourt between 17.05 and 17.12 but his luck had then run out. His Spad was hit by ground fire from Leutnant Mangel's BZ35 defence unit and he was wounded in the chest. He tried to reach the front lines but was forced down in a field outside the village of Murvaux. He had come down 50 yards from a small stream. Climbing out of his machine, he tried to crawl to the water, it must be assumed to drink, as the type of wound he received induces a raging thirst. It was also getting dark and German troops were looking for him. The popular story that emerged was that he fired at the soldiers and was then shot and killed, as if defiant till the end. What most probably occurred was that he fired his pistol in order to attract attention to himself so that he might have medical attention, but died as the troops got to him. In any event he was one of the most well known fliers of the USAS to emerge from the war, not only for his amazing run of 18 victories in just over two weeks, but also because he was the only US pursuit pilot to receive the Medal of Honor in the war.*

October 1918
Christian Kairies, who had shot down Bill Craig on 26 September, was himself shot down on the first day of this month. No.210 Squadron's Camels were coming back from patrol, and over Houthulst Forest at 3,000 feet they were attacked by 12 Fokker DVIIs. Lieutenant C W Payton DFC (E4421) was hit in the pressure tank so rolled over and down, low over the Forest while turning on his gravity tank. He had been hit by Kairies, others seeing the Camel go down and later reporting that their missing comrade had at least bagged a Camel before being shot down himself. However, it seems that as Payton sorted himself out, he then saw two Fokkers chasing another Camel, one of which was Kairies again. Payton turned, closed in and fired at close range at one Fokker and felt he had hit the pilot by the way the Fokker went down. Payton was then engaged by two more Germans but got away from them. Meanwhile, Captain E Swale and another pilot had seen the Fokker crash into the Forest below.

MFJV lost just one pilot, Kairies. He came down inside the British lines badly wounded and died the next day. The 'victory' claimed on his behalf was recorded as his seventh. It was Clement Payton's tenth victory. He would gain his 11th and last later in the day, although 210 Squadron would suffer the loss of its CO and another pilot in a fight with MFJIII. The very next day Payton was killed by a direct hit from AA fire whilst bombing a train near Courtrai.

---

* Eddie Rickenbacker also received the Medal of Honor for his war service but it was not awarded until 1930.

SEPTEMBER – OCTOBER 1918

Fritz Höhn

A former soldier with the 7th Guards Infantry Regiment, Fritz Höhn came from Wiesbaden and was 22. Flying two-seaters prior to becoming a fighter pilot he had scored his 21st victory on the third day of October and might, perhaps, have wondered about the possibility of receiving the Blue Max. However, his next combat this same day ended everything.

He had fought with Jastas 21s, 81 and 60 prior to taking command of Jasta 41 just a couple of days earlier. At least ten of his victories had been balloons and virtually all his successes had been achieved on the French Front. He was shot down near St Martin l'Heureux, evidently by pilots of GC12 – the Storks. It is not certain who fired the fatal shots, only that Lieutenant Robert Le Petit of Spa67 shot down a German between Dontrien and Moronvillers and Sergent Robert Brillaud of Spa26, one west of Somme-Py. St Martin is about ten kilometres west of Somme-Py and one kilometre south of Dontrien. Both claims were timed at 14.30.

Young John Warner

John Warner was 19, from Boston Spa, Yorkshire. He claimed eight victories, was awarded the DFC and then killed in action, all within a space of three and a half months. He flew with 85 Squadron, and a Fokker pilot of Jasta Boelcke got him on 4 October. Paul Bäumer and Gerhard Bassenge were the claimants, shooting down SEs over Montbrehain and Joncourt. They were their 40th and sixth victories. The action occurred at noon, and three SEs of the Squadron were lost. Jasta 36's Vizefeldwebel Alfred Hubner claimed an SE5 near Le Catelet, just to the northwest, time unknown, which might have been the third – even Warner. It was his sixth victory too.

Adjutant Pierre Delage, from the Dordogne, had been in the infantry too prior to becoming a pilot. However, he was 31, but had scored seven victories with Spa 93, won the Medaille Militaire, had been made a Chevalier de la Légion d'Honneur as well as receiving the Croix de Guerre with Palmes. He was lost at 11.45 near Machault north-east of Reims but it is difficult to discover his fate.

Two aces also fell to ground fire in this first week of October: Captain M L Cooper DFC of 213 Squadron (six victories), downed by machine-gun fire as he strafed a train on the 2nd (F3951), and Unteroffizier Hermann Marwede of Jasta 67, downed by gunfire as he went for a balloon of the American 6th Balloon Company. He had just flamed his fourth balloon for victory number five but failed to make it back over the lines, and he was captured. The event was dramatically, and uniquely, recorded by an American cameraman.

Herbert Boy's lucky day

It is much easier to learn what happened to Leutnant Herbert Boy on 7 October 1918. Boy had been a Schusta pilot, gaining combat experience in these fighting two-seaters before going on to fighters. With Jasta 14 he had served long and hard, achieving five victories.

On 7 October, his Jasta met up with a three-Flight patrol of 29 Squadron's SE5s, in layers, which included a number of successful pilots, one being a French Canadian, Captain Camille Henri Raoul Lagesse DFC. (His comrades rather ungraciously, but with humour, had a nickname for him, corrupting his family name to suggest an excess of posterior!) He would end the war with 20 victories and a Bar to his DFC, as well as the Belgian Croix de Guerre.

At 08.45, leading the top five SEs between Staden and Courtrai, Lagesse spotted seven Fokker biplanes diving on one of the lower formations. As he and his men edged towards them, six of the Fokker pilots, obviously seeing them, turned east, but one continued to pursue the lower bunch of SE5 machines. Lagesse dived on this aircraft, firing 250 rounds. The Fokker caught fire and went down, the pilot being seen to jump over the side and open his parachute. As he watched, the French-Canadian saw the luckless airman's clothes on fire and then the parachute too, low to the ground. As Lagesse circled, the man hit the ground, did not move while flames continued to rise from the crumpled mass of man and parachute. Returning to base, having also shared the destruction of a two-seater with another pilot, Lagesse reported the action, assuming that the German Fokker pilot had burned to death. However, Herbert Boy survived. He had come down on the Allied side and was taken to hospital to have his injuries treated. Luck had twice been with him this day.

The next day, another American serving with the RAF, Lieutenant Cleo Francis Pineau from Albuquerque, New Mexico, gained his eighth victory but was brought down on the German side of the lines to become a prisoner. At 09.23, he and his 210 Camel Squadron companions were four miles east of Roulers at 11,000 feet and met 11 Fokker fighters, biplanes and Triplanes! (Although the DVII biplane had all but equipped every German Jasta, some Triplanes continued to be used, some pilots, such as the ace Josef Jacobs, actually preferring the machine to the DVII.) Pineau, in D1868, was seen by Captain Ed Swale and another pilot to shoot down one Fokker DVII, his sixth victory and his sixth DVII. However, he was then hit and brought down. When news of his capture came through, the award of the DFC was pursued and later approved. Another 210 pilot also failed to return.

The other pilots of 210 also claimed a further three Fokkers destroyed, one in flames and two crashed. Although the record of claims by the German Jasta pilots in the final weeks of the war are incomplete, it looks like 210 Squadron were in combat with Jasta 51, who claimed three Camels near Roulers. One was claimed by Leutnant Karl Plauth, the Staffelführer, his 13th victory, the other by Leutnant Carl Berr, his first (although he had claimed four unconfirmed victories as a two-seater pilot). However, Berr was also shot down in the fight and slightly wounded, having to bale out. He landed badly and was severely injured, but survived. As the claims by the other pilots were all for DVIIs, it can be assumed that Plauth was flying a Triplane, and so it was he who brought down the American. Plauth would end the war with 17 victories.

Paul Bäumer makes it 43
By early October, Leutnant Bäumer had scored over 40 air combat victories, a list that had commenced with Jasta 5 in July 1917 and had risen spectacularly with Jasta Boelcke. He was one of the last to receive the Pour le Mérite and had been lucky enough to survive a bale-out in September.

On 9 October he added his final victory to his impressive tally. No.62 Squadron's Brisfits were escorting a bomb raid and over the Forêt de Mormal ran into trouble. Jasta 2 hit them soon after crossing the lines, having taken off at 06.45. The fight lasted to Aulnoye Junction and back to the lines, two BF2b machines being shot down, and another shot up with one crewman wounded. In one of the downed machines (E2256) was Captain Lyn Campbell, and observer Second Lieutenant W Hodgkinson. They had last been seen going down over the Forest, under control, but with two enemy fighters on their tail. The second lost crew were taken prisoner, but Campbell and Bill Hodgkinson were both killed. Campbell, from Hamilton, Ontario had gained nine victories. It was Bäumer's final kill.

Wilbert White's sad loss
Studying and researching men's activities in wars could make one hardened to the waste and suffering but most students will probably admit to moments of sadness when reading of some events, even if they occurred long before they themselves had even been born. One such story for me was the death of Lieutenant Wilbert Wallace White Jr, a pilot in the 147th US Aero Squadron.

Many men try to dodge fighting in a war, and who can blame them, but there are also the men who perhaps could legitimately avoid it but who don't. I believe one such was White. On 1 May 1918, this New

Yorker had already reached his 29th birthday, was a married man with a son and a daughter and was working in a publishing company. Nevertheless, he decided where his duty lay and had enlisted into the USAS in July 1917. He eventually found himself in France with the 147th. By late September 1918 he had become an ace, won the DSC and French Croix de Guerre. Early in October he was notified that he could return to America for a rest, and would probably not have returned to France before the war had ended. However, like so many others before and since, he made the fateful decision to fly 'just one more sortie'.

This was on 10 October, a day he could easily have been packing his bags before heading for the ship home. Instead he led a patrol over the Verdun Sector around noon and shot down a two-seater for his seventh victory, shared with his Flight. That too could have been it, but the afternoon saw another patrol flown, this time to attack a balloon near Doulcan, so White led that too. Over Dun-sur-Meuse the Spads ran into Jasta 10 and a dog-fight began. One will never know exactly what occurred or what was in the American's mind in his last few moments. However, from the reports of the other pilots, it seems that his guns had jammed and in seeing one of his men under pressure from a Fokker, he deliberately rammed the German. He saved his fellow countryman and achieved an eighth victory but at the cost of his own life. The German pilot, Leutnant Wilhelm Kohlbach, however, had the means to escape and took it. As Wilbur (as he was usually called) fell to his death in Spad 7588, Kohlbach was able to take to his parachute and survive. The German also received credit for the Spad, which brought his score to five.

White's actions brought forth a recommendation for the Medal of Honor but in the event he only received a posthumous Oak Leaf Cluster to his DSC. Little recompense for his widow and two young children and even the story of his heroic end would never replace a loving husband and father.

14 October 1918
Two British and one American aces came to grief on this autumn day. The first was Lieutenant C M Wilson of 29 Squadron. This Canadian from Vancouver, who had previously been in the artillery, joined the RFC in Canada and was flying with 29 Squadron by the late spring of 1918. With eight victories achieved over the summer he had received the DFC, but his luck ran out on this day. Flying SE5a F5516, he had been part of a 29 Squadron patrol which took off at 07.45 led by Captain Lagesse. The six pilots ran into Fokker DVIIs near Roulers and engaged in combat at 08.50. Lagesse landed at La Lovie with his radiator shot through at 09.00 while another landed at 7 Squadron's base with a holed fuel tank. Another

had been forced to abort before the fight started and two more were wounded but got back at around 09.15, one reporting a hit by AA fire. Claude Melnot Wilson was last seen east of Roulers in combat.

Although Lagesse claimed two Fokkers in the fight, one of which came down inside British lines to be captured, there is no indication as to whom the pilot was; certainly there is no note of any German fighter pilot being taken prisoner, so one must assume he did not survive the action. A Jasta 63 pilot claimed an SE5 this day, over Ypres, time not known, and Jasta 43 claimed four in the same locality, although the fourth claim was given to the Jasta 63 pilot. There is also a suggestion that at least one may have been a Camel (west of Roulers at 17.00) so these claims may be in doubt. Unteroffizier Siegfried Westphal of Jasta 29 claimed two SEs this day, victories five and six, but place and times are both unrecorded. This Jasta was operating from Aertrycke on the German 4th Army Front, so the sector is correct. Leutnant August Burkhard of this unit also claimed an SE5 on this day for his fifth victory. It is perhaps relevant, that Jasta 29's records are very sparse after the end of September, so they may also have had a pilot lost to Lagesse. The only other SE5 losses on the 14th were from 2 AFC Squadron, with two pilots failing to get back, one killed, one POW, near Tournai, which is far too south to be either Jasta 43 or 63's fights.

Howard Clayton Knotts, a 23-year-old pilot from Illinois, was an ace with the 17th US Aero, but was shot down and taken prisoner this day. Flying on an OP over the British 3rd Army Front in D3328, the 17th Aero had taken off at 12.55. The 15-strong Camel formation attacked ground targets on the Verchain Road and in the villages of Monchaux and Quérenaing, then south of Vendegies. However, either hit by ground fire or with an engine problem, Knotts was forced to land inside the German lines. He was seen to get out of his Camel but was soon taken prisoner.

With six victories, Knotts had won the British DFC and American DSC, and while being shipped to a prison camp by rail he had somehow managed to set fire to a freight train which was carrying aircraft to the Front, which resulted in seven Fokkers being destroyed. Perhaps his score should read 13? He was nearly put in front of a firing squad but survived to become an aviation lawyer, dying of a heart attack in 1942, aged 47.

John Greene's loss
Captain John Edmund Greene DFC came from Winnipeg and was 24. He had joined the RNAS in 1916, like so many of his countrymen, and had flown with 13 Naval Squadron from late 1917 until the unit became 213 Squadron RAF in April 1918. Becoming a flight commander he had

fought throughout the summer and no doubt was feeling tired by October.

This may have been reflected on 4 October, for that day he got himself shot down by Carl Degelow of Jasta 40. In this fight Greene had claimed three victories, and Jasta 40 did lose one pilot, but two of the Camels had been hit. Greene, in one B7270, had got down safely on the right side of the lines although another pilot was killed.

Ten days later, Greene had another busy day. In the morning he claimed his 15th victory but that afternoon 213 were in a fight with 17 Fokkers over Pervyse at 14.30 and lost three Camels. Greene, in D3409, was one of the missing. The other two were dead, one being an American US Navy pilot, Lieutenant Ken MacLeish. Greene crashed at La Panne and died soon afterwards from a fractured skull in Pervyse hospital, and was buried in the Coxyde Franco-British Cemetery.

The opposition had been the Marine pilots from MFJIV. Reinhold Poss had claimed two, one over Houthulst Forest another over Zarren (victories nine and ten), while Flugmeister Gerhard Hubrich had scored his ninth kill with a Camel over Iseghem.

CHAPTER FIFTEEN

October – November 1918

Vizefeldwebel Albert Haussmann was six days past his 26th birthday on 16 October. Serving with Jasta 13 since the summer, he already had an impressive record. His first victory had been achieved with Kasta 26 two years earlier, and then, with Jastas 23 and 15, and now 13, his score had reached 15. On this day he was shot down by ground fire near Romagne on the Ardennes Front, and although he attempted to bale out, his parachute had no time to open before he hit the ground. Most of his victories had been scored on the French and American sectors.

An own goal
There is no definitive list of aircraft shot down by their own side in WWI but there are a number known to historians. One I record here occurred on 18 October, concerning the German ace Leutnant Oliver Freiherr von Beaulieu-Marconnay, who had scored 25 victories with Jasta 15 and then as commander of Jasta 19. From Berlin–Charlottenburg, he was just 20 years old.

A former Dragoon cadet he had seen action in the east and had been commissioned at 17 prior to becoming a pilot. 'Bauli' as he was known, scored his first kill in May 1918 and was now in line for the Blue Max. However, in an air battle with some Spads on this day, it is understood he was fired on by a Fokker from Jasta 74. Badly wounded in the upper thigh over Gouzecourt he crash-landed and was taken to a hospital at Avion. It was fairly obvious the young Staffelführer wasn't going to make it so his Pour le Mérite was rushed through and he was told of his high honour shortly before he died on the 28th.

The Marines bag two aces
On 23 October, 204 Squadron lost five Camels and their pilots. They had taken off at 08.30 to fly a High OP after dropping some bombs on front-line targets. One pilot had suffered engine trouble on take-off and he eventually crash-landed at Furnes although he himself was unhurt. The patrol ran into twelve aircraft from MFJII, clouds and mist allowing the Germans to hit them before they knew what was happening. Two of the Camels were also seen to collide, but all were claimed by the Marine

pilots; in fact they actually claimed seven, with Oberleutnant zur See Gotthard Sachsenberg, CO of MFJI, claiming two (victories 27 and 28), Vizeflugmeisters Alexandre Zenses three (16 to 18), Karl Scharon (seventh) and Leutnant Merz one each. The two aces of 204 lost were Captain T W Nash DFC (D9608) and Lieutenant O J Orr (D9613). Tom Nash, a Sussex lad aged 26, eight victories, and 23-year-old Osborne Orr, from Cleveland, Ohio, USA, aged 23, died. The other three were also killed.

Bréguets get Klaudet

Jasta 15's six-victory ace Vizefeldwebel Gustav Klaudit from eastern Prussia and a former Uhlan scored his victories in 1918. All were over the French and American Sectors and it was the Americans that finally put him out of the war. Attacking Bréguets of the 96th Aero Squadron on 23 October he was wounded in the left arm as his Fokker received a burst of machine-gun fire over the Bois de Money. The war ended while he was still having treatment. Seven crews of the 96th Aero, led this mission by Captain B F Beverly, soon to become CO of the 100th Aero, claimed a Fokker shot down over Landres at 16.00 hours. However, he may also have been hit by a Spad in the same action.

Schlegal goes down fighting

Vizefeldwebel Karl Schlegel of Jasta 45, a 25-year-old Saxon, had been a former machine-gunner pre-war and had then seen action in France and Russia with the 8th Cavalry Division. Upon becoming a pilot in 1916 he had flown two-seaters and been a home defence fighter pilot prior to being assigned to Jasta 45 in May 1918. He was also one of the balloon specialists, and of his 22 victories, 14 had been 'drachens'. On 27 October he sortied across the lines to attack balloons once again, and in fact frontline witnesses seem to indicate a balloon and a two-seater Spad shot down at La Malmaison at 15.40 (G) on his last sortie although these appear not to have been officially confirmed.

Schlegal didn't get back, being engaged by French Spads and killed, falling near Amifontaine on the French side. Three Fokkers were shot down some distance north-east of Amifontaine this day but it is thought he had been shot down by the French ace Sous Lieutenant Pierre Marinovitch of Spa94 (down over Le Thour) at 14.05. It was the Frenchman's 19th victory of an eventual total of 21, scored in just over a year. He had won the Medaille Militaire and been made a Chevalier de la Légion d'Honneur, plus having the Croix de Guerre with ten Palmes. He was to die in a flying accident on 2 October 1919.

Raesch gets another

We have met Josef Raesch before. He had shot down Captain E C Hoy of 29 Squadron, then survived a bale-out after being set on fire by Hind of 40 Squadron. His score had now reached five but he made it six on 27 October with another ace from 29 Squadron.

Captain Guy Wilbraham Wareing was only 19 but he had been made flight commander and won the DFC with this SE5a-equipped fighter unit. Since August he had claimed five aircraft and flamed four balloons. However, the lad from Warrington, flying SE5 H676, met up with Jasta 43 this day and Raesch shot him down east of Tournai at 10.30 (G), Wareing falling to his death.

The 29 Squadron patrol had been to the south of Tournai area and at 09.20, Wareing and Lieutenant S M Brown had attacked a balloon just to the east of the town, which caught fire after Brown's attack. The SEs had then been attacked by ten Fokker biplanes and one monoplane (a DVIII?), and Camels from another squadron had joined in. One Fokker had been seen to go down out of control. At 09.50 seven Fokkers dived on SE5s of yet another squadron south of Renaix. The patrol engaged these too and two Fokkers were claimed destroyed. However, the patrol leader did not emerge from these combats.

The Fokkers were noted as having white tails and carrying a black bar on the fuselage, very like the Jasta 43 markings. Josef Raesch would gain one more kill on the 30th to bring his score to seven.

Barker – VC

There are a number of air actions that seem so well reported that it is difficult to admit little is really known of what went on. A certain Canadian's VC action in June 1917 is a case in point, and another Canadian won his VC in an action which is well reported but lacking in many details. Unlike the June 1917 action, we at least know Captain William George Barker was in combat, if for no other reason than his multiple wounds.

Billy Barker already had the DSO and Bar, MC and two Bars, Croix de Guerre and the Italian Gold Medal for his work in France and Italy over two years. This 25-year-old Manitoban had been an observer, two-seat pilot, fighter pilot and squadron commander, scoring some 46 victories in that time. Returning to England after much success in Italy against the Austro-Hungarians, he could also have seen out the war, but was still eager to get back into the fighting. Finally he managed to persuade people that if they indeed wanted him to instruct, he should at least know something of the air combat scene in France at that time, so he was given a period of ten days attached to 201 Squadron, taking with him one of the new Sopwith Snipes to evaluate. However he failed to get into any combats until 27 October, but on this day, over Mormal Wood (actually flying home to England), he found, engaged and shot down a Rumpler two-seater at 08.25. However, he was then attacked by a dozen or more Fokker DVIIs, and in a spirited fight was severely wounded in both thighs and one elbow. With each wound he lost consciousness but each time he came round to fire at his antagonists. He finally crash-landed in Allied lines, south of Valenciennes, the fight having taken place in full view of front-line troops, and witnesses had stated seeing anything up to ten German aircraft brought down. Sifting through the reports, and from what little Barker himself could remember, he was given credit for three Fokkers, which brought his score to 50, but his injuries put him out of the war, although it brought him the award of the Victoria Cross.

It would be nice to record instantly with whom he had been engaged but the identity of the German unit involved has so far eluded historians. It would not be uncharitable to say there were probably not as many German fighters as Barker himself thought, for he probably imagined that each time he came round, he was fighting other Germans, and of course soldiers often exaggerated what they saw, or didn't totally understand what they were seeing. From what we know today, only some five German fighter pilots are shown as casualties this day and one of those was Schlegal down on the French Front, and two others were taken prisoner on this Front too. A fifth pilot was wounded on the St Quentin part of the front. Nor is there a nice Snipe claim, but this might not be significant as it would probably have been mistaken for a Camel, and coming down on the British side, it might not have received confirmation. Of the four 'known' Camels claimed this day, two were by the Marine pilots over Belgium, another two far south on the German 2nd Army Front and the last also in Belgium. All the claims over Belgium were undoubtedly losses sustained by 203 and 204 Squadrons.

Karl Schoen Jr

From Indianapolis, Indiana, Lieutenant Karl Schoen Jr flew with the 139th Aero in 1918 and claimed seven victories. He was killed in combat over Damvillers on 29 October. His last two victories were credited from reports of others in his final fight, timed at 15.20. It is not easy trying to find the German unit, but Jasta 6 claimed two Spads between 16.25 and 16.30, which is the right time, although we don't know the location.

Callender and Farquhar

The pilot of Jasta Boelcke knocked down two British aces on 30 October.

Captain A A Callender, a New Orleans American, has been mentioned before. He led a patrol of 32 Squadron out at 08.25 for an OP to the Ghislain area in E6010 and was last seen in action there an hour later. They ran into aircraft of Jasta Boelcke and had three men killed and another taken prisoner. Alvin Callender, with eight victories, was one. The other ace was Lieutenant R W Farquhar (D6132), another experienced air fighter. He had flown FEs with 18 Squadron in 1916, and then Spads with 19 Squadron the following year. He was actually claimed by Manfred von Richthofen in June 1917, but survived to fight another day. From Dulwich, he had gained one last victory with 32 to bring his score to seven but had fallen in this action. Jasta Boelcke's Ernst Bormann (his 15th), Lindenberger (11th) and Oberleutnant von Greisheim (first) had each claimed one SE5, at Harchies, north of Neuville and at Fresnes.

First man in Ostend
Lieutenant Archie Buchanan, a 26-year-old American from Long Island, New York, was the first member of the Allied forces to enter Ostend. That occurred on 17 October when he landed his Camel there and was greeted with the news that the Germans had left that morning.

Buchanan had been with 210 Squadron since June but had scored seven victories and won the DFC. However, fame, such as it was, came to an end on 30 October. Flying Camel F3242 on a morning Line Patrol he was last seen east of Valenciennes in combat with seven Fokkers at 11.15. Fortunately he survived the encounter as a prisoner, his period of incarceration being the last 12 days of the war. The only Camel claimed in this area was that shot down by Unteroffizier Michael Hutterer of Jasta 23b, over Sebourg at 12.20 German time, and Sebourg is just east of Valenciennes. No.210 claimed three Fokkers this day, Jasta 23b losing one pilot. Hutterer was a Bavarian from Munich and had been serving his country since 1914, first with a machine-gun company, then the artillery and finally the air service. Since May 1918 he had scored six victories, Buchanan being the seventh. His eighth and last came on 1 November.

Other American aces downed this day were with the USAS 22nd Aero, 1st Lieutenants R deB Vernam and James Dudley Beane. Vernam too came from New York, having volunteered to fly with the Lafayette Flying Corps and serving with Spa96, with which he had claimed one victory. This 22-year-old had transferred to the USAS, assigned to the 22nd Aero and raised his score to six and won the DSC. Beane came from New York as well and had brought his score to six the previous day.

Missing from patrol on the 30th, Vernam's Spad XIII was claimed by

Leutnant Bertling of Jasta 12, his first victory. Remington Vernam came down inside German lines, badly wounded in the groin. With the Germans in retreat, he was abandoned in a hospital at Longwy and found by the Red Cross. By this time he was seriously ill and he succumbed on 1 December. Beane's Spad, No.7812 '14', crashed near Betheville. He was the fifth and final victory of Vizefeldwebel Otto Klaiber of the same Jasta, the combat timed at 16.30. Bertling survived the war, having scored two more victories in the last days of the war, and Klaiber too saw out the war, having been a two-seater pilot back in 1917.

The last hurrah!
Sunday 3 November, just a week to go before the guns fell silent, but few would have guessed it. For all anyone knew the war would continue into 1919. This day, however, saw Leutnant Heinrich Maushake's last battle. Having scored six victories with Jasta 4 over the last five months, this Circus pilot is believed to have run foul of the US 103rd Aero and been severely wounded. The fight was timed at 16.50 (G) around Stenay and Andevanne, three Spads being claimed by other pilots of the Jasta. However, if it was an American unit, it was more likely to have been the 22nd Aero, who did have three of its pilots shot down south-west of Beaumont-en-Argonne, and one of its pilots, 1st Lieutenant Harmon C Rorison, claimed three victories; was one of these Maushake? The area is right.

To die in any war is an abomination. To die in the last days of a war is an abomination and an immense pity. In the last days of the first war in the air, the fighting was as intense as it had been at any time. The Germans, lacking fuel and numbers, nevertheless put up a stout resistance till the end. This was particularly so on 4 November. No.65 and 204 Squadrons, for instance, were engaged in a massive air battle with the Marine pilots south-east of Ghent on this morning.

They had taken on a large force of Fokkers over Melle, right on the Schelde River, south-east of Ghent, at 08.45. No.204 estimated the force to be around 20 to 38 German fighters sporting red noses and white tails, while 65 Squadron noted Fokkers with yellow and green tails, chequered and light fuselages. In this fight claims for German aircraft shot down far exceeded the known losses, even accepting the possible shortage of German records at this late stage. However, Jasta 29 had a fight with Camels this day and claimed five, although the time and location is not known. All we do know is they were at Aertrycke on the German 4th Army Front. There were any number of Camels lost this day, so it would be a complete guess to say who shot down whom.

OCTOBER – NOVEMBER 1918

Certainly MFJIV were involved, Flugmeister Gerhard Hubrich scoring his two final kills of the war, his 11th and 12th, in the Ghent area. No.65 Squadron lost two with another pilot wounded; 204 lost two also, one being the ace Lieutenant John Douglas Lightbody (F6257). This Scot, from Hamilton, had celebrated his 19th birthday just five days earlier, having joined the Squadron in mid-September. In October he had claimed five victories. Purely as a matter of interest, 65 Squadron claimed eight destroyed, six out of control (and one driven down), while 204 claimed two destroyed and five out of control – 21 victories! If Jasta 29 were in this action, they lost one pilot.

The Australians take a hammering
No. 4 AFC Squadron had exchanged its Sopwith Camels for the new Sopwith Snipe fighters in October and had already started to score well with them. However, on this fourth day of November, they ran into Jasta Boelcke and lost three, two of the pilots being aces.

These two were Captain T C R Baker DFC MM (E8065) and Lieutenant A J Palliser (E8064). The Australians had taken off at 11.40 and the three missing men had last been seen in combat with Fokkers east of Buissenal at 13.15. Ernst Bormann claimed one, Karl Bolle two, east of Renaix, near Tournai and over Escanaffles. Tom Baker was buried at Escanaffles; he had scored 12 victories and had won the MM and Bar whilst an NCO with the artillery in 1916. Arthur Palliser, from Tasmania, had seven victories. It had been a really bad day for 4 AFC, for additionally they had lost two pilots in the morning, both taken prisoner, and they too had been shot down by Jasta Boelcke, and Bolle had got them both, making it four Snipes for him bringing his final war score to 36. Bormann's kill was also his last, making a war total of 16.

The end
When the SE5s of 29 Squadron landed back at base on the morning of Saturday 9 November 1918, the bad news was that two pilots were missing. One was the Canadian, Second Lieutenant W A Howden (E5999) who was killed, the other was the ace Lieutenant E O Amm (F853). Later, the good news was that Amm had survived his last combat.

The Squadron had flown out at 08.55 to patrol between Munte and the front lines. Fights with Fokker DVIIs had started at 09.45 which continued on and off till 10.30. First came a scrap with 15 DVIIs east of Audenarde in which one German machine was seen to crash. This was followed by a battle with seven Fokkers and three two-seaters to the north-east of this town. A Flight's SEs destroyed two Fokkers, possibly another, and crashed a two-seater. Several German soldiers came out to one of the crashed and burning Fokkers, two SE pilots shooting them up. Then, south-east of Ghent, more DVIIs were encountered; two crashed near Laerne and another went down out of control. Meanwhile, B Flight had attacked seven DVIIs of the first 15 fought, Amm getting an out of control victory and another pilot destroying one which was seen to crash. Six more Fokkers were fought south-east of Ghent and another was sent down out of control.

Amm and Howden continued to fight over Lemberge, where Amm saw his companion shoot down a Fokker in flames but he was immediately brought down by two fighters. Amm also destroyed one in flames but being attacked by three more he had his controls shot away and he spun down. He managed to right his SE at 500 feet but it then became uncontrollable again and he crashed into some trees, thereby escaping serious injury. The three enemy fighters continued to strafe him while still in the tree but he was not hit and he later managed to extricate himself but was quickly taken prisoner by German troops. He was locked up in Antwerp but as the German evacuated the city he was left behind, and on 19 November 1918 he arrived back at the Squadron on a push bike. With his own two claims and recording one for Howden, it made the score nine destroyed and seven out of control, plus one forced to land, for the two lost SEs.

There is no obvious record as to which Jasta or Jastas they were fighting, which is a pity – it was probably too near the final collapse to have records completed let alone survive, and certainly it would be difficult for the system to have claims confirmed. As far as is known, only one German pilot was lost this day and that was a Jasta 65 pilot taken prisoner, but that was down on the French Front.

At 11.00 hours on the morning of 11 November 1918, the guns were silenced and peace came to the Western Front. The Great War, and the first air war, was over. The aeroplanes and the men who flew them had come a long way since the first faltering steps back in August 1914. Future wars would never be the same again. The aeroplane was here to stay.

EPILOGUE

If nothing else, one might say the aces of both sides were the élite among their peers, or at least, managed to score more victories than most – or, in some cases, managed to persuade 'higher authority' that they had. I don't mean this to sound facetious or derogatory, but it is now well known that overclaiming did go on, and this was due in part to the system and the way the war was fought mainly over the German lines; the British in particular claimed far more than could ever have actually been shot down.

However, leaving that aside, there were still a goodly number of very interesting men who scored well and who undoubtedly fared better in the actual destruction of the opposition than some others. It is perhaps a waste of time to speculate on how the Services of both Britain and Germany might have been different in later years had some of them survived. But it is interesting.

It seems to be a fact that most flat-out combat pilots do not achieve high office in peacetime, and that most commanders in the RAF after WWI (and even WWII) were often those who had organisational abilities rather than fighting abilities. Or perhaps they had the perceptibility to 'go with the flow' and not 'rock the boat'. Perhaps this is how it should be, or is it? Assuming some of the fallen aces had even wanted to continue in their country's service, one wonders how men like James McCudden VC would have fitted in, or Mick Mannock, Albert Ball, Rod Dallas, Lanoe Hawker; or Manfred von Richthofen, Werner Voss, Rudolf Berthold, Oswald Boelcke, Lothar von Richthofen.

Is it foolish to dream of scenarios? McCudden would probably have made a good Air Chief Marshal, or even Marshal of the Royal Air Force, provided his record had out weighed the fact that he was the son of an army corporal and had himself started life as a lowly bugler. In the first quarter of the twentieth century such things mattered in some circles. However, he was a thinking man, a planner, and someone who could see the advantages of certain courses of action. Mannock was a killer, an out-and-out combat pilot who, while well educated and studious, might also have suffered due to being the son of a corporal. Whether or not he would have received the VC had he lived is also speculation. One imagines that he would not have wanted to stay in the RAF; too much red tape and too many bureaucrats for his liking.

Albert Ball, VC or not, what might he have achieved had he continued flying and fighting? Perhaps he would eventually have been brought down and captured, for he was happiest when flying alone and those times had almost gone when he was killed. If he had survived the war, he might well have gone into the aircraft design field – he had already made a start with the Austin-Ball Scout. With the influence of his father behind him, a Mayor of Nottingham and later knighted, and money, he might well have been persuaded to become a local figurehead within industry. He disliked the killing so may well have welcomed the peace with its opportunities of doing other, more worthwhile things.

Roderick Dallas could well have become something in the Service, perhaps well up in the Australian Air Force, or perhaps he might have gone into the Fleet Air Arm and commanded an aircraft carrier. That might well have suited his character and ambitions.

Lanoe Hawker had the right attitude, the right background as the son of a Lieutenant RN, and had himself been commissioned into the Royal Engineers after being a cadet at the Royal Military Academy. With his engineering background he may well have progressed within the technical side of the Service, although one feels he would have had to fly and test things himself. He was a doer, with a hands on approach, but he would certainly have received the respect his record would have given him, even demanded.

The von Richthofen brothers? What might Manfred have achieved had he lived? One could imagine him either gaining high office in the German Air Force or equally retiring to the family estate to hunt and be a guest of the rich and famous around the world. Would Adolf Hitler have approached him to lead his Luftwaffe when he came to power in the early 1930s? Would Richthofen have refused due to his background and knowing full well what type of man Hitler was? Would he have tried to oppose the Nazi regime? How would he have fared against Göring? Would he have failed against both? As for brother Lothar, he would probably have been killed in a plane crash just as he was. Flying was more in his blood, but he pressed too hard.

Voss? How would his Jewish background have survived in Nazi Germany. Another high-scoring Jewish pilot who died in the war had his name actually removed from the list of air heroes at one stage, but Voss seemed to be too well known and respected. But perhaps like Ball, he would not have wished to continue in a service life, but might well have done well in aviation. His flying ability was too remarkable to be lost. He also had that adventurous streak that might have led him to long-distance flying, helping to open up the air routes of the world in the 1930s. And then, might the problem of living with the Nazi regime perhaps have forced him to leave Germany altogether?

Rudolf Berthold was a dedicated German and a fighter, but with an air of common sense, despite his apparent disregard for personal safety. He too may have tried to oppose a future regime with which he would probably have disagreed, or would he have seen some glimmer of hope in restoring Germany's good name, only to find, too late, that it was in reality a sham? Boelcke was another that must have been a rival for Göring's job had Manfred von Richthofen refused. He would probably have smiled that his protégé had been offered it first, always assuming that Boelcke's score had not kept ahead of Richthofen's. Perhaps Boelcke might have ended 1918 with 150 victories, while Manfred could only reach 110!? But had Richthofen ended up as top scorer, and with perhaps just that little bit more charisma away from the Service scene, one would doubt equally that someone with the brains of Boelcke would have had much to do with Hitler and his party. One could only hope that he would not have ended up like Udet and been used as a puppet and given things to do that were on the one hand beyond him and on the other against his nature.

What, one wonders, might the world be like had some of these men not fallen in the Great War, but then that could be said of many thousands who gave their lives in this most terrible conflict. Perhaps if men like these, having survived, could have formed a sort of Nation's League of Heroes there wouldn't have been a second Great War.

INDEX

Adam, Oblt H 51, 52, 88-9
Adams, 2/Lt H T 114
Alexander, Lt T M 171
Alexander, FL W M 46
Allen, 2/Lt C M 193
Allmenröder, Ltn K 35, 46-7
Allmenröder, Ltn W 35, 36
Altmeier, OffStv F 50, 195
Amm, Lt E O 209-210
Anderson, Lt G B 29
Anderson, Lt R W L 44
Andrews, Capt J O 6, 10, 11
Armstrong, Lt F C 29
Arnison, Lt C H 128
Aspinall, Capt J V 126
Auffarth, Oblt H 154
Auger, Capt A V R 38
Avery, Lt W 158

Baer, Lt P F 132-3
Baier, Ltn K 136
Bailey, Lt G G 118
Baker, 2/Lt F G 84
Baker, Lt H E 22
Baker, Capt T C R 209
Baldamus, Ltn H 28
Ball, Capt A 14, 32-4, 50, 211-2
Bamberger, S/Lt 162
Banks, Capt C C 118
Barcat, S/Lt A J L 152
Barker, Maj W G 205
Barksdale, Lt 171
Barlow, Capt L M 60-1
Barnekow, Ltn R von 124
Barton, Lt D H 127, 145
Bassenge, Ltn G 198
Bäumer, Ltn P 195, 197, 199
Baurenfeind, Ltn K 129
Baux, Adj A 152-3
Bayley, Lt F G 183
Baylies, Lt F L 146
Beamish, Capt H F 29
Beane, 1/Lt J D 207
Beanlands, Capt B P G 12-13, 102, 109
Beaulieu-Marconnay, Ltn O von 203
Becker, Ltn H 14, 45, 123
Beckmann, Ltn L 155, 156, 179
Belgrave, Capt J D 143-4
Bell, SC B C 46
Bell, Capt D J 133-4
Benbow, Capt E L 136, 194
Benger, Sgt W J 85
Bennett, Lt L Jr 173
Bennett, Lt R G 135-6
Bennett, Lt R M 183
Bennett, Lt S L 29
Bennetts, FSL R S 67
Benzler, Ltn A 103
Berr, Ltn C 199
Berry, 2/Lt O W 23
Berthold, Hptm R 3, 82, 83, 98-99, 111, 165, 177, 211, 212-3
Bertling, Ltn 208
Bertrab, Ltn J von 65-6
Besemüller, Vfw A 100

Bethge, Oblt H 67, 103
Beverly, Capt B F 204
Biddle, Maj C J v
Bill, Lt A G 49
Billik, Ltn P 31, 114, 130, 139, 151, 164
Birkbeck, 2/Lt R A 84
Bishop, Maj W A 158
Blake, Lt A G S 156
Bleibtreu, Ltn 103
Blum, Ltn 191
Blumenbach, Oblt P 102
Blumenthal, OffStv F 167
Boddien, Oblt H-H von 193
Boelcke, Hptm O v, 1, 3, 8-9, 10, 11, 16, 21, 91, 156, 211, 213
Boenigk, Oblt O von 56, 71, 77
Boger, Capt W A 165
Böhme, Ltn E 38, 91, 92
Böhner, Ltn 40, 68
Böhning, Ltn H 109, 185
Bolle, Rittm K 140, 148, 175, 209
Bond, Capt W A 57-8
Bongartz, Ltn H 52, 87, 90, 121-2
Böning, Ltn W 137-8, 178
Booker, Maj C D 63, 64-5, 80-1, 90, 168-9
Borm, Uffz G 192
Bormann, Ltn E 185, 207, 209
Bossler, Vflugm 42
Bosson, Sgt A L 153
Boulton, Lt N S 194-5
Boumphrey, Capt G 18
Bowen, 2/Lt C S 115
Bowman, Maj G H 55, 79, 82, 86
Boy, Ltn H 198
Boye, Sgt 35
Boyou, Lt M J-P 182
Bozon-Verduraz, S/Lt J 73
Brandt, Ltn F 185
Brauneck, Ltn O 58, 60-2
Breadner, Capt L 29, 69-70
Bretschneider-Bodemer, Ltn M-W 153
Brillard, Sgt R 197
Britnell, FSL F J S 99, 185
Britton, Lt A F 48-9
Broad, Capt H S 29
Broadberry, Capt E W 50-1
Brocard, Maj F 38
Brockhoff, Ltn z S 108
Brookes, Capt E G 163
Brown, Capt A J 176
Brown, Capt F E 100, 104
Brown, 2/Lt J R 152
Brown, Lt S M 205
Brown, Lt W H 104
Brunnecker, Sgt B 154
Bryers, 2/Lt G 20
Buchanan, Lt A 207
Büchner, Ltn F 150
Buckstett, Ltn 109
Buckler, Ltn J 91
Buder, Ltn E 153
Buddecke, Oblt H-J 98-9
Bullen, 2/Lt E H 155

Bülow-Bothkamp, Ltn H von 123-4
Bülow-Bothkamp, Ltn W von 92, 93, 124
Burge, Capt P S 155-6
Burger, Lt M G 163
Burkhard, Ltn A 201
Bush, Lt J C 83
Busk, Lt C W 12
Busse, Ltn J von 85

Cairnes, Capt W J 139
Caldwell Maj K L 45
Callaghan, Maj J C 149, 150, 174
Callender, Capt A A 151, 164, 206-7
Campbell, Capt L 199
Campbell, Capt W T 44
Carlin, Capt S 186
Carpenter, Lt P 88
Carter, Maj A D 89, 129-30
Carter, Capt A W 29
Carter, Lt R B 68
Case, 2/Lt C H 194-5
Casey, Lt F D 29
Chapman, Lt V C 137
Chapman, Lt W W 83
Chaput, Lt J 110, 123
Chester, Lt 133
Chick, Capt J S 126
Chidlaw-Roberts, Capt R L 77-8, 94
Chisam, 2/Lt J R 183
Chisam, FL W 69
Christian, Lt W L 111-2
Clark, Capt C C 123
Clark, Lt N 97
Clarke, Lt E D 86
Clarke, Lt H A 128
Classen, Ltn E 181
Clauditz, Ltn 144
Claxton, Lt W G 170-1
Claydon, Capt A 151
Claye, Lt C G 34-5
Claye, Capt H 101, 128
Clayson, Capt P J 144
Clement, Capt C M 40, 67-8
Cochrane-Patrick, Capt W J C K 36
Cock, Capt G H 58-9
Cockerall, Lt S 22
Coghill, Lt W H 185
Coghlan, Lt 133
Collett, Capt C F 70-2
Collier, Lt J 45
Collin, Ltn D 133, 156, 169
Collins, Lt E W 128
Collins, 2/Lt V H 19
Collishaw, Maj R 46, 88
Colville-Jones, Lt R 121
Colville-Jones, Capt T 121
Coombes, Capt L P 146-7
Cooper, Capt M L 160-1, 197
Cordonnier, S/Lt A P J 160
Cour, Lt P V de la 126
Court, Sgt L S 13
Cowell, Sgt J J 161
Cowie, 2/Lt W A 178
Cowper, Lt A K 102
Cox, Lt G M 88-9

INDEX

Craig, Lt W B 183, 187-8, 196
Crane, 2/Lt S W 125
Crawford, Capt K 10, 117
Creutzmann, Ltn O 144
Crole, Capt G B 91
Cruise, 2/Lt M G 185
Crummey, Lt F C 191
Crossen, Lt E B 141-2
Crow, Capt C M 33-4
Crozet, Cpl 35
Crundell, FSL E D 63
Cubbon, Capt F R 42, 43
Cudemore, 2/Lt C W 63
Cull, Capt A T 36
Culling, FSL T G 42
Cummings, Lt H W M 169-170
Cummings, 2/Lt L 90
Cunnell, Capt D E 49
Curphey, Capt W G S 37-8
Curtis, Lt R L 76-7
Cymera, Vfw W 14-15, 35

Dahlmann, Ltn T 178
Dahm, Ltn W 53, 87
Dallas, Maj R S 42, 138-9, 211-2
Daniel, Lt H C 118, 119
Dannhuber, Ltn X 82, 84
Dawe, Lt J J 141-2
Davidson, Lt W D 126
Davies, Lt H E 27
Davies, 2/Lt L G 40
Day, FL H 96
Day, FC M J G 97
Debect, Lt 21
Debenitz, Vfw 133
Degelow, Ltn C 152, 164, 177-8
Deilmann, Ltn K 52, 67
Delage, Adj P 197
Delling, Ltn A 128
Delorme, S/Lt A J 16
Demisch, Ltn M 179, 187
Dennett, Lt P M 139-140
Derode, Capt J M E 140
Deullin, Capt A L 38
Devilder, S/Lt 172
Dicksee, Lt H J H 14
Dickson, Lt J L 30
Dietlen, Ltn A 117-8
Dilthey, Ltn H 152
Dodds, Lt R E 69
Dolan, Lt H E 122, 124
Donaldson, Lt J O 154, 164, 177-8
Döring, Oblt K-B von 43-4
Dorme, Lt R P M 13, 38-9
Dossenbach, Ltn A 13, 38, 48, 63
Dostler, Oblt E von 49, 51, 52, 68-9, 88
Douillet, S/Lt 148
Doumer, Capt R 21
Doyle, Capt J E 120-1, 180
Doyle, Capt 133
Drake, Capt E B 195
Dran, 2/Lt A C 114
Drekmann, Ltn H 161-2
Dreyer, Flugm 97
Dubonnet, MdL P M 146
Dunton, Lt 133
Durrad, Capt F A 87
Durrant, Capt T 127
Dutrou, S/Lt 93

Eaton, Lt E C 147
Edwards, Capt C G 175

Ehrlich, Lt J L 183-4
Ehrmann, Vfw F 118-9
Elder-Hearn, Lt T 90
Ellwood, FL A B 99
Emsden, Cpl L 27
Engelfried, Flugm K 188
Esswein, OffStv O 96, 153
Evans, 2/Lt H C 7, 8
Ewers, Oblt W 124-5, 132
Eyre, FC C A 50

Fahlbusch, Ltn W 12
Fall, Capt J S T 29
Farquhar, Lt R W 207
Farrell, Lt C M G 163
Ferreira, Lt H M 52, 97
Fervers, Uffz K 195
Festler, Ltn M 163
Festner, Vfw S 28, 30
Filbig, Ltn 21
Findlay, Capt L 18
Finney, 1AM F 121
Firth, Capt J C B 59, 88-9
Fitzgibbon, FL D F 46, 76
Fitton, 2/Lt J C 126
Flecken, Ltn G W 54
Flemming, Uffz 19
Fletcher, Sgt R M 192
Flynn, Capt H L W 164
Foley, Lt R G 98
Fonck, Lt R P v, 38, 81
Foote, Capt E L 17 fn
Forman, Capt J H 178-9
Foster, Lt G B 141-2
Foster, Capt R M 167
Fouquet, Sgt 172
Fowles, 2/Lt 168-9
Frankl, Ltn W 3, 22-3, 26
Frantz, Ltn 97
Friedrichs, Ltn F 148
Frommherz, Ltn H 140, 175, 178, 181
Fruhner, Ltn O 96, 184-5
Fry, Capt W M 45, 93-4
Fullard, Capt P F 53, 54, 90
Fullerlove, 2/Lt G Y 52-3

Galley, Capt E D G 120
Gallwitz, Ltn K 86, 113
Gandert, Oblt H-E 136, 194
Gardner, Capt C V 166, 192
Garros, Lt R 1
Gebhardt, Gfr 104
Geigl, Ltn H 115
Gerstenkorn, Uffz 123
Gilbertson, Lt D H S 137-8, 178-9
Gildemeister, Ltn J 171
Gilmour, Capt J 132
Gilroy, Lt E C 126
Glover, 2/Lt C L 126
Glynn, Capt C B 122
Gonne, Capt M E 163
Good, Lt H B 180
Goode, Lt C A 12
Goodison, 2/Lt F B 23
Gontermann, Ltn H 35
Gordon, Lt D 22
Gordon-Davis, 2/Lt C 17
Göring, Oblt H W 20, 55, 62, 77, 143, 152, 212-3
Gorringe, 2/Lt F C 93-4
Got, S/Lt 172
Göttsch, Ltn W 17, 47, 80, 116-7
Graham, 2/Lt J 69

Grassmann, Ltn J 160, 166
Grandmaison, Capt D L 35
Gray, Lt J 98
Gray, Lt W E 161
Greene, Capt J E 201-2
Gregor, Vfw K 154
Greim, Oblt R von 128
Greisheim, Oblt von 207
Grimm, Ltn 160
Gullmann, Uffz 162
Guynemer, Capt G M L J v, 13, 38, 73-4, 85

Haase, Gfr 123
Haber, Ltn K 13
Haebler, Ltn H G von 83-4, 108
Haegelen, S/Lt C M 110
Hahn, Oblt E 31, 70
Hale, Capt F L 192
Hales, Capt J P 173
Hall, Capt D S 83
Hall, Lt R W P 14
Hamersley, Capt H A 78
Hamilton, Lt L A 133, 173
Hammes, Ltn K 71-2
Hand, Lt E M 88-9
Hanning, Lt J E 111-2
Hanstein, Ltn L 71-2, 107
Hantelmann, Ltn G von 145, 182, 184
Harker, FSL M A 96
Harker, Lt 168
Harrison, 2/Lt N V 80
Harrison, Lt T S 141
Harrison, Lt V W 13
Harrison, Capt W L 57
Hartigan, Lt E P 83
Hartley, 2/Lt J H 59
Hartmann, Oblt O 69-70, 71
Hartney, Maj H E 18-19
Hasdenteufel, S/Lt M 146
Hauss, MdL M 21
Haussmann, Vfw A 203
Hawker, Maj L v, 1, 10-12, 13, 117, 211
Hay, Lt R B 56
Hazel, AM J P 152
Hechler, Flgr 175
Hedley, Capt J H 113
Heinemann, Uffz W 4
Heinrich, Ltn z S B 39, 108, 176, 179
Heins, Ltn R 114
Heinz, Ltn 175
Heinzelmann, Ltn W 134
Heldmann, Ltn A 166
Helmigk, Ltn H 61
Henderson, Lt I 60
Henke, Ltn F 142-3
Hennrich, Vfw O 193
Hertz, Ltn K 115
Hess, Ltn E 53, 67, 92
Heurtaux, Capt A M J 7, 13, 38, 76
Hewat, Lt R A 170
Hickey, Capt C R R 183, 187-8
Hilf, Ltn W 22
Hill, Lt A H 98
Hill, Lt C W 161
Hill, Lt R F 80
Hills, Pbr J 161, 170
Himmer, Ltn 180
Hines, Sgt G F 191
Hind, Lt I F 157, 168, 205
Hobson, Lt P K 100, 104
Hodgkinson, 2/Lt W 199

Hoffmann, Ltn P 102
Höhn, Ltn F 197
Höhndorf, Ltn W 3, 7
Hoidge, Capt R T C 50, 79, 81-2
Holland, 2/Lt C B 23, 26
Holliday, Capt F P 36
Holme, Capt R C L 141
Holmes, Lt D S 27
Hopkins, 2/Lt 17
Howard, Capt R W 108-9
Howarth, Capt N 176
Howden, 2/Lt W A 209-210
Hoy, Lt E C 193, 205
Hoyer, Ltn H 89-90
Hubbard, Lt W 133
Hübner, Vfw A 197
Hubrich, Flugm G 202, 209
Hudson, Capt F N 53-4
Hughes, Capt G F 101, 128
Hughes, Lt W 129
Hunter, Lt D Y 150
Hutterer, Uffz M 207

Imelmann, Ltn H 16-17
Immelmann, Oblt M 1, 3, 6, 10, 16
Inglis, Lt D C 158-9
Irving, Capt G B 166
Irwin, Capt W R 120, 165

Jacobs, Ltn J C P 69, 72, 198
Jailler, Adj L J 35
James, Lt L R 195
Jameson, 2/Lt J 36
Jardine, Lt R G 55, 62
Jarvis, Capt L 127
Jeckert, Uffz L 179
Jeffs, Lt C H 83
Jenkin, Lt L F 39-40, 44, 75-6
Jenkins, Lt W S 176, 179
Jenks, Lt A N 41
Jenson, Ltn J 191
Joens, Uffz 144
Johns, Capt W E 5
Johnson, Lt G O 104, 145
Johnson, Lt O P 180
Johnston, Capt J E 92
Johnston, FSL P A 67
Jones, Capt H W G 21-2
Jones, Capt J I T 124
Jordan, FSL W L 65
Joslyn, Lt H W 18, 48, 67
Jourdan, 2/Lt W T 19
Junor, Capt K 120, 127, 180

Kahlow, Vfw M 128
Kairies, Flugm C 188, 196
Kaleta, Uffz 114
Kampe, OffStv W 97-8
Kauffmann, Uffz E 119
Kelly, Lt E T S 144
Kemsley, Capt N 93
Kennard, Lt W D 17
Kennedy, 2/Lt D 115
Kermedy, Lt H A 57-8
Kensen, Vfw P 195
Kent, FSL R 46
Kerby, Capt H S 29
Kergoloy, Adj De 92
Keudell, Ltn H von 19-20
Kieckhafer, Ltn F 85
King, 2/Lt C F 114, 118
Kirk, 2/Lt P G 52-3
Kirkham, Capt K R 113

Kirschbaum, Uffz 132
Kirschstein, Ltn H 126, 127-8
Kirmaier, Oblt S 9-10
Kister, Vfw 172-3
Kleffel, Ltn W 83
Klein, Ltn H 28, 35, 96-7
Kindley, Lt F E 147
Kitto, Capt F M 115
Klaiber, Vfw O 208
Klaudet, Vfw G 204
Klimke, Ltn R 175, 185-6
Klingman, Lt E C 177
Knake, Ltn B 59
Knight, Capt A G 13-14
Knight, Lt D 136
Knight, Lt G 186
Knobel, Vfw 175
Knotts, Lt H C 201
Knowles, Capt R M 17
Kockeritz, Ltn F von 176
Koepsch, Ltn E 120-1, 180
Kolbach, Ltn W 200
König, Ltn E 22
Könnecke, Ltn O 117, 121, 128, 141, 168
Kosmahl, OffStv F 71, 77-8
Krebs, Vfw F 52, 54-5
Kressner, Vfw K 121
Kroll, Ltn H C 39, 62, 104, 170
Kruger, Ltn R 50
Kuhl, Ltn 13
Kuhlmann, Uffz 87
Kuhn, Ltn M 95
Kunst, Ltn O 70
Kunstler, Flugm 40
Küppers, Ltn K 51

Lahoulle, Lt A 110
Lagesse, Capt C H R 141, 198, 200-1
Lale Capt H P 161
Lambert, Capt W C 144, 145
Landis, Capt R G 154
Lane, Lt W I E 116-7
Langan-Byrne, Lt P A 8
Langlands, Lt D 59
Larsen, Lt J F 104
Latimer, Capt D E 172-3
Latta, Capt J D 43
Lavers, Lt C S I 54
Lawson, Capt G E B 191
Leake, Lt E G 112
Leask, Capt K 104
Leed, Lt D K 167
Leete, Lt S J 49
Leffers, Ltn G 3, 14
Lehmann, Ltn W 148
Leitch, Capt A A 147
Lenoir, Adj M A 9
LePetit Lt R 197
Leptien, Ltn H 170
Lesée, Cpl 21
Letts, Lt J H T 35-6
Lewis, Lt R G 143
Lewis, Lt T A M S 41
Leycester, 2/Lt P W 91
Lightbody, Lt J D 209
Lindeberg, Lt E W 145
Lindenberger, Ltn A 207
Lindsay, 2/Lt A T W 97
Little, Capt R A 119, 134
Lloyd, Lt D C R 45
Lloyd-Evans, Lt D 155

Loerzer, Hptm B 83, 143, 153, 175, 178, 181
Loerzer, Ltn F 143
Longton, 2/Lt E S W 30
Louden, Capt L G 114
Lowe, Capt C N 141
Löwenhardt, Ltn E 97, 140, 160, 164-5
Loyd, Capt A T 82
Luchford, Capt H G E 17, 80, 92, 93
Lufbery, Maj G R 131-2
Luke, Lt F Jr 184, 195-6
Lutjöhann, Ltn W 125
Lux Vfw A 178, 181

MacAndrew, 2/Lt C 83
MacDonald, Lt E R 120
Macdonald, Lt R M 194
MacGregor, FL N M 76
Maclanachan, Lt W 57
MacLeish, Lt K 202
MacMillan, Capt N C 75
McCall, Capt F R 143, 167, 170-1
McCormack, Lt P F 183
McCubbin, Lt G R 4
McCudden, Lt J A 104-5, 127
McCudden, Maj J T B v, 65, 79, 82, 104, 211
McDonald, Capt I D R 141, 145
McDonald, Lt R 124
McElroy, Capt G E H 162, 173
McGinnis, Lt J 98
McGregor, Capt M C 165-6
MacKay, Capt A E 93
McKilligan, 2/Lt F F 191
McMaking, Lt O L 75
McMillan, Capt J C 17
McQuiston, Capt F 166

Maasdorp, 2/Lt C R 114
Maashoff, Ltn O 36
Madill, Lt R McK 41
Madon Lt G F 70, 153
Magoun, Lt F P 116
Mai, Ltn J 126, 141, 192, 194
Makepeace, Lt R M 48
Malcolm, Lt R G 27
Malone, FSL J J 29, 31
Mangan, Lt D C 174-5
Mangel, Ltn 196
Manley, 2/Lt P S 191
Mannock, Maj E v, 65-6, 122, 124, 158-9, 162, 211
Mansbridge, Lt L M 39-40
Manschott, Vfw F 21
Mansell, Lt W S 75
Manteufel, Ltn W von 107
Manzer, Capt R 163
Mappin, Lt K G 98
Mare-Montembault, Capt M J J G 21
Marinovitch, S/Lt P 204
Mark, Lt R T 102
Market, Vfw G 137-8
Marquardt, Vfw O 51
Marsden, 2/Lt C 114
Marty, Lt A P L M 142
Marwede, Uffz H 197
Matton, Capt J G F 72
Maushake, Ltn H 208
Maxwell, Capt G J C 60-1, 82
May, Capt W R 150, 167, 185-6
Maybery, Capt R 79, 92
Mealing, Lt M E 110-1
Meggett, Lt W G 87

INDEX

Meintjes, Capt H 32-4
Mellings, Capt H T 154-5
Menckhoff, Ltn C 79, 82, 158
Mendel, Lt K A 140
Merz, Ltn 204
Mesch, Vfw C 181
Messervy, Capt E D 51
Mettlich, Oblt z S K 43, 100, 104
Meyer, Ltn G 166, 180
Miller, Lt A W B 52
Millward, FSL K H 50
Milne, 2/Lt C G 185
Milne, Capt J T 85
Mindner, Ltn H 97
Minifie, FL R P 104
Minot, Capt L 48-9, 63
Mohr, Ltn A 22
Moissinac, MdL X J-M L 140
Montrion, Adj R 148
Montgomery, Capt K B 88-9
Moore, 2/Lt E S 36
Moore, Capt G B 116
Moore, 2/Lt M C 59
Morse, Lt T W 86
Mortemart, S/Lt F DeR de 103
Mottershead, F/Sgt T 17
Mouldre, S/Lt O P de 123
Müller, Ltn H 111-2, 116, 133
Müller, Ltn H K 14
Müller, Ltn M 94
Mülller, Uffz 21
Mulzer, Lt M von 4, 5
Murphy, Sgt P 126
Muth, Vfw A 61-2

Napier, Capt C G D 126
Nash, FL G E 46
Nash, Capt T W 204
Nathanael, OffStv E 36
Näther, Ltn M 193
Neckel, Ltn U 105, 150, 169
Negben, Ltn W 172-3
Neuenhofen, Vfw W 154
Newland, Sgt A 172
Nicholson, Lt W E 91
Niederhoff, Ltn A 50, 56, 59, 62
Noel, Lt T C 172
Noltenius, Ltn F 186
Nungesser, Lt C E J M 13
Nuville, S/Lt C M J L 146

Oberlander, Ltn H 133
O'Brien, Lt C R 30
O'Callaghan, 2/Lt M A 68-9
Oesterreicher, Uffz A 4
Ogden, 2/Lt 90
Oliver, Capt T A 66
Olivier, Lt E 89
Olley, Sgt G P 44
Olsen, Ltn W 22
O'Neill, 2/Lt 114
Orchard, FSL W E 29
Orlebar, Capt A H 101
Orr, Lt O J 204
Osten, Ltn G von 115
Osterkamp, Oblt z S T 51
Osterroht, Hptm P H A T von 28-9
Owen, Lt R J 114-5

Page, FL J A 58
Palliser, Lt A J 209
Palmer, Lt 145
Parry, Lt C 120

Parschau, Ltn O 3, 6
Partington, Lt O J 56
Patey, Capt H A 176, 179-80
Patrick, Capt W D 116
Pattern, 2/Lt J A 91
Payton, Lt C W 196
Peacock, 2/Lt F C 129
Pech, Vfw K 131
Pègoud, Lt A C 1, 2, 3
Peiler, 2/Lt M F 114
Pentland, Capt A A N D 173-5
Perkins, Lt J F R I 98
Petit, Adj P A 184
Phillips, Capt R C 143
Piercey, Sgt W 7
Pinder, Capt J W 160-1
Pineau, Lt C F 198
Pippart, Ltn H 152, 166-7, 169
Plange, Ltn R 129
Platel, 1AM S H 86
Plauth, Ltn K 171, 199
Potter, Pvt F A 18, 48
Pratt, 2/Lt S H 20
Poeze, Lt de la 162
Poss, Ltn z S R 161, 202
Pourpe, Lt M 132
Powell, Maj F J 95
Prehn, Vfw A 4
Pressentin, Ltn V von 138
Preuss, Ltn W 140
Price, FC G W 96
Price, 2/Lt W T 34-5
Prior, 2/Lt W J 114
Prothero, Capt P B 51, 60-1
Purcell, Lt J 161
Putnam, Lt D E 181-2
Pütter, Ltn F 128, 184

Quandt, Ltn T 85, 177-8
Quested, Capt J B 14
Quigley, Lt F G 93-4

Rackett, Lt A R 109
Raesch, Ltn J 157, 168, 194, 205
Rahn, Ltn A 142
Ramsey, FSL D W 50
Randell, 2/Lt A C 37
Raymond-Barker, Maj R 119
Rayner, Lt J W 180
Rautter, Ltn V von 114
Redler, Lt H B 57, 102
Reed, Lt A E 141
Reeves, 2/Lt H G 84
Reid, FL E V 46, 63
Reinau, Ltn R 146
Reinhardt, Oblt W 59
Rhys Davids, Lt A P F 79, 82, 85-6, 113
Richards, Lt C R 49, 67
Richardson. Lt H B 103
Richardson, Capt L L 27-8
Richthofen, Ltn L S von 27, 33-4, 35, 36, 101, 128, 167-8, 211-2
Richthofen, Rittm M A von v, 10-12, 13-14, 18-19, 26, 27, 28, 33, 44, 46, 48, 59, 66, 67, 76, 78, 101, 105, 112, 116, 118-9, 122, 136, 150, 152, 164, 195, 207, 211-3
Rickenbacker, Capt E V 196 fn
Riessinger, Vfw R 45
Ritzinger, Uffz 87
Robert, Adj M J E 153
Robin, S/Lt 148

Robinson, 2/Lt C F 193
Robinson, 2AM T N O 4
Robinson, Capt W L 26
Robson, Lt C C 106-7, 126
Rochford, Capt L H 179
Rochford, Lt S W 179
Rogers, Lt W W 44
Rolfe, Lt R B 140
Rooper, Lt W T V 44, 84
Rorison, 1/Lt H C 208
Rose, 2/Lt F 56
Rosenau, Gfr 134
Rosenbachs, Ltn 13
Rosencrantz, Ltn H 12
Rosenfeld, Vfw O 150
Ross, Capt C G 141
Röth, Oblt F 161
Roy, Lt I L 153-4
Roulstone, Capt A 103
Rumey, Ltn F 114, 132, 141-2, 147, 191-2
Rumpel, Ltn T 96
Runge, Ltn R 83, 88
Rushden, Capt C O 139

Sachsenberg, Oblt z S G 39, 204
Saimer, Sgt 148
Salter, Lt G T C 136
Sanday, Maj W D S 12
Sanderson, Lt I 146-7
Sandys-Winsch, Lt A E 177
Satchell, Capt H L 41
Saundby, Capt R H M S 11
Sauvage, Sgt P J 16
Savage, Lt D A 128-9
Savage, Lt J R B 4, 5
Saward, Lt N C 71
Sawden, Lt W W 41
Schäfer, Ltn H 144-5
Schäfer, Ltn K E 27, 40-2, 144
Scharon, Vflugm K 204
Scheele, Ltn F C von 37
Scheibe, Ltn 132
Schickler, Ltn H 169
Schilling, Oblt H 13
Schlegel, Vfw K 204, 206
Schlimpen, Gfr J 171-2
Schmidt, Oblt O 76, 193
Schmitt, Ltn R 157
Schobinger, Ltn V 65, 86, 90
Schoen, Lt K Jr 206
Scholtz, Ltn E 123
Schönfelder, Obflugm K 70, 140, 146
Schrottt, Vfw 9
Schulte, Ltn A 27
Schulte-Frohlinde, Ltn F 176
Schultz, Ltn H 141
Schulz, Ltn W 140, 146
Schumm, Uffz M 155-6, 170
Schuster, Ltn G 96
Schwarz, Ltn K 153
Selig, Ltn 123
Sellers, Lt H W 106-7, 126
Selwyn, Lt W E 144, 145
Sewell, Capt S 150
Sharman, FC J E 57
Sharpe, Lt F 44
Sharpe, Capt T S 119
Sharples, Lt N 68-9
Shepherd, Capt A S 56
Shields, Capt W E 171
Shook, FC A M 60 fn, 108
Siddall, Lt J H 156

Sidney, 2/Lt L P 83
Simon, Cpl 18
Sloley, Lt R H 60, 82
Smith, Lt D E 170
Smith, Lt E A L 85-6
Smith, 2/Lt E F W 14
Smith, Lt G S 187
Smith, 2/Lt N 183
Smith, Capt S P 116
Smith, 2/Lt W C 53
Sneath, Lt W H 115
Soden, Capt F O 94
Somerville, 2/Lt H A 94-5
Southey, Lt J H 141, 145
Spille, Ltn 117
Stead, Lt I O 178
Steinhäuser, Ltn W 148
Stephens, Lt E J 171
Stephenson, Sgt T F 86
Stephenson, Capt W S 159-60
Stock, Lt E R 126
Stock, Ltn W 49
Strange, Capt G J 187
Strange, Col L A 187
Strähle, Ltn P 19, 174, 190-1
Strasser, Vfw G 53
Stubbs, 2/Lt C L 100
Summers, Capt J K 167-8
Swale, Capt E 196, 198
Swayze, Lt W K 179
Symons, Lt H L 89

Taplin, Lt L T E 181
Taylor, Capt A G V 85
Taylor, Lt E 173
Taylor, Lt E J 176
Taylor, 2/Lt F J 19
Taylor, Lt H L 116-7
Taylor, Lt M S 123, 150
Taylor, Capt StC C 38, 104
Tegedor, Ltn 110
Thayre, Capt F J 42, 43
Theiller, Ltn R 22
Thomas, Ltn E 109-110
Thomas, Lt 12
Thompson, Lt C W M 195
Thompson, Lt M L 172
Thompson, Capt S F 192
Thompson, 2/Lt S F 67
Tidmarsh, Capt D M 23, 26
Tilney, Maj L 139
Tipton, 1/Lt W D 175

Tissier, Sold. 93
Tod, 2/Lt G D 125, 163-4
Todd, Capt J 137-8
Todd, 2/Lt R M 175
Tolman, 2/Lt C 193
Tonks, Capt A J B 169-70
Trapp, 2/Lt D 88
Trapp, FL G L 58, 87-8
Trapp, FSL S 88
Travers, Lt H G 29
Trescowthick, Lt N 181
Trollope, Capt J L 113-5
Trotsky, OffStv J 124
Trusler, Lt J N 154
Trusson, 1AM A 36
Tudhope, Capt J H 57, 63, 81
Tuffield, Lt T C S 69
Turnbull, Lt E 60
Turnbull, Lt J S 51
Tutschek, Oblt A von 20-1, 63-5, 101-2
Tuxen, Ltn R 49
Tyrrell, Capt W E 142

Udet, Oblt E v, 114, 138, 168, 188-9, 213
Ulmer, Ltn A 17-18, 47-8
Unger, Ltn H 110
Unger, Lt K 146-7
Uniacke, 2/Lt D P F 76-7
Upfill, Lt T H 112
Urquhart, 2/Lt A 67

Veitch, 2/Lt C L 30
Veltjens, Oblt J 7, 165
Venmore, 2/Lt W C 103
Vernham, 1/Lt R deB 207-8
Viebig, Ltn H 148
Violet-Marty, Adj P A F 15
Voigt, Oblt B von 43
Voss, Ltn W 66, 75, 76, 78-9, 86, 211-2

Waddington, Lt M W 48
Wagner, Ltn R 84
Waldhausen, Oblt H 80-2
Walker, 2AM J C 27
Walker, Lt K M 167-8
Walkerdine, Lt H J 110
Wall, Capt A H W 36
Waller, Cpl J H 4-5
Walton, Lt O T 27
Walz, Hptm F J 38

Wareing, Capt G W 205
Warner, Lt J W 197
Warman, Capt C W 62
Watson, Lt K B 137-8, 178-9
Wear, 2/Lt A E 49
Webb, Capt W W N 60-2, 66
Wedel, Lt E von 126, 176
Wehner, Lt J F 184, 195
Weigand, Oblt E 66
Weir, FSL C H 58
Weiss, Ltn H 122-3
Weissner, Ltn E 42
Weightman, 2/Lt H 71
Wells, Lt C D 128
Wendelmuth, Ltn R 80
Wentz, Ltn A 165
Werner, Vfw A 56
Werner, Ltn J 139
Wessels, Ltn s S H 182-3, 187
Westphal, Uffz S 186, 201
Whealy, Capt A T 29, 99
White, Capt H G 130-1
White, Lt V R S 80, 92
White, 1/Lt W W 199-203
Whitehead, Capt L W 125, 132
Whittaker, Lt H A 141
Williams, Lt R 175
Wilson, Lt C M 200-1
Wilson, Lt E B 145
Windisch, Ltn R 135
Winkler, Lt W O B 35
Winter, FC R R 95-6
Winterbotham, Lt F W 52
Wingtens, Ltn K 3, 7, 10
Wise, FSL S E 108
Wissemann, Ltn K 73-4, 81
Wolff, Ltn H 105, 126-7
Wolff, Oblt K 26, 50, 76
Wollen, 2/Lt D C 27-8
Wood, Lt H H 168
Woodbridge, 2/Lt A E 49
Woollett, Capt H W 118
Wright, Lt S 85
Wright, Lt 38
Wüsthoff, Oblt K 56, 145-6

Yerex, Lt L 179
Young, Capt W E 45, 59

Zenses, Vflugm A 204
Zetter, Uffz 110
Zimmer, Capt G F W 94-5